The Holobiont Imperative

Thomas C. G. Bosch • David J. Miller

The Holobiont Imperative

Perspectives from Early Emerging Animals

Thomas C. G. Bosch
Zoological Institute
Christian Albrechts Universität zu Kiel
Kiel
Germany

David J. Miller
ARC Cnt. of Execl. for Coral Reef Stud.
James Cook University
Townsville, Queensland
Australia

ISBN 978-3-7091-1894-8   ISBN 978-3-7091-1896-2 (eBook)
DOI 10.1007/978-3-7091-1896-2

Library of Congress Control Number: 2016931994

Springer Wien Heidelberg New York Dordrecht London
© Springer-Verlag Wien 2016
This work is subject to copyright. All rights are reserved by the Publisher, whether the whole or part of the material is concerned, specifically the rights of translation, reprinting, reuse of illustrations, recitation, broadcasting, reproduction on microfilms or in any other physical way, and transmission or information storage and retrieval, electronic adaptation, computer software, or by similar or dissimilar methodology now known or hereafter developed.
The use of general descriptive names, registered names, trademarks, service marks, etc. in this publication does not imply, even in the absence of a specific statement, that such names are exempt from the relevant protective laws and regulations and therefore free for general use.
The publisher, the authors and the editors are safe to assume that the advice and information in this book are believed to be true and accurate at the date of publication. Neither the publisher nor the authors or the editors give a warranty, express or implied, with respect to the material contained herein or for any errors or omissions that may have been made.

Printed on acid-free paper

Springer-Verlag GmbH Wien is part of Springer Science+Business Media (www.springer.com)

Acknowledgments

No matter how many authors are on the cover, every book is the work of many hands. We would like to thank the many people who made this project possible, starting with the contributions of many students, postdocs, and co-investigators over the years. TB thanks particularly to René Augustin and Sebastian Fraune, who initiated work on innate immunity and host-microbe interactions in *Hydra*. The work related to this review was supported in part by grants from the Deutsche Forschungsgemeinschaft (DFG) and the Clusters of Excellence "The Future Ocean" and "Inflammation at Interfaces" (to TB). DJM gratefully acknowledges the support of the Australian Research Council, both directly and via the ARC Centre of Excellence in Coral Reef Studies, and the unlimited patience of Eldon Ball, a real scholar and remarkable human being.

Contents

1 Introduction: The Holobiont Imperative 1
   1.1 Of Complex Diseases and Animals as Complex Systems:
       Why Bacteria Matter 2
   1.2 The Complexity of Coevolved Animal Communities
       Was Discovered in 1877 in Kiel, Germany 6
   1.3 Looking for a Term for the Functional Entity Formed
       by a Host and Its Associated Microbial Symbionts............... 8
   References.. 9

**2 Major Events in the Evolution of Planet Earth:
Some Origin Stories** 11
   2.1 Microbes Were First: Bacteria Have Existed
       from Very Early in the History of Life on Earth 11
   2.2 Life Did Not Take Over the Globe by Combat,
       But by Networking... 11
   2.3 The Transformation of the Biosphere
       at the Ediacaran–Cambrian Boundary........................ 13
   2.4 Our Bacterial Ancestry Is Reflected
       in Our Genomic Signature.................................. 15
   2.5 Genomes of Early Emerging Metazoans, Similar
       to Humans, Contain a Considerable Fraction
       of Genes Encoding Proteins of Bacterial Origin 16
   2.6 The CRISPR/CAS System as Window
       into Ancient Holobionts..................................... 19
   2.7 Origins of Complexity: What Makes an Animal?................ 20
   2.8 Multicellularity Requires Cooperation of Cells.................. 21
   2.9 Genomes of Early Emerging Metazoans Reveal
       the Origin of Animal-Specific Genes.......................... 22
   References... 25

**3 The Diversity of Animal Life: Introduction
to Early Emerging Metazoans** 27
   3.1 How Old Are the "Early Diverging" Animal Phyla?.............. 28
   3.2 Cnidarians: The Closest Relatives of "Higher"
       Animals (Bilateria)... 32

	3.3	Sponges: One Phylum or More?	34
	3.4	The "Comb Jellies": The Enigmatic Phylum Ctenophora	39
	3.5	Placozoans: The Simplest Extant Animals?	40
	3.6	Eyes, Nervous Systems, and Muscles	42
	3.7	The Closest Unicellular Relatives of Extant Animals	43
	3.8	The Paucity of Data on Symbioses Involving "Lower" Animals	44
		References	45
4	**Phylosymbiosis: Novel Genomic Approaches Discover the Holobiont**		47
	4.1	Animal Life and Fitness Is Fundamental Multiorganismal	47
	4.2	Phylosymbiosis and Coevolution	48
	4.3	Microbiota Diversification Within a Phylogenetic Framework of Hosts: Insights from *Hydra*	51
		References	54
5	**Negotiations Between Early Evolving Animals and Symbionts**		57
	5.1	Cnidaria Use a Variety of Molecular Pathways to Elicit Complex Immune Responses	57
	5.2	How Do Cnidarians Distinguish Between Friends and Foes: Insights from Corals and *Hydra*?	60
	5.3	Selection Can Favor the Establishment of Mutualisms and Animal–Microbe Cooperation	61
	5.4	Rethinking the Role of Immunity	63
		Conclusion	64
		References	65
6	**Role of Symbionts in Evolutionary Processes**		67
	6.1	Microbes as the Forgotten Organ	67
	6.2	Developmental Symbiosis	68
	6.3	The Role of Symbionts in Evolutionary Processes	69
	6.4	*Nematostella*, an Early Metazoan Model to Understand Consequences of Host–Microbe Interactions for Rapid Adaptation of a Holobiont to Changing Environmental Conditions	71
	6.5	Rapid Adaptation to Changing Environmental Conditions: The Coral Probiotic Hypothesis	73
	6.6	The Role of Symbionts in Speciation	75
		References	76
7	**The *Hydra* Holobiont: A Tale of Several Symbiotic Lineages**		79
	7.1	Rationale for Studying Host–Microbe Interactions in *Hydra*	80
	7.2	The Hydra Microbiota	82
	7.3	Linking Tissue Homeostasis, Development, and the Microbiota	84

7.4	Hydra's Mucus Layer Plays a Key Role in Maintaining the Necessary Spatial Host–Microbial Segregation..............	86
7.5	Microbes Differ in Embryos and Adult: Embryo Protection.......	87
7.6	Antimicrobial Peptides Function as Host-Derived Regulators of Microbial Colonization.........................	91
7.7	Symbiotic Interactions Between *Hydra* and the Unicellular Algae *Chlorella*	93
Conclusion		95
References..		96

8 Corals... 99

8.1	The Case of Reef Building Corals: A Complex Association Between Animal, Algal, and Bacterial Components.................................	99
8.2	Attempts to Generalize About Coral–Microbe Interactions Are Complicated by the Evolutionary and Physiological Diversity of Corals...	100
8.3	The Complexity of Coral Microbial Communities	101
8.4	Where Are the Bacteria Located?	102
8.5	Transmission Mode and Ontogeny	103
8.6	Key Components of the Coral Microbiome...................	104
8.7	Nitrogen-Fixing Bacteria Are Intimately Associated with Corals..	107
8.8	Probiotic Microbes and Antimicrobial Peptides	107
8.9	Coral–Bacterial Interactions Modulate Local Climate Via Sulfur Metabolites....................................	108
Conclusion		109
References..		110

9 Bleaching as an Obvious Dysbiosis in Corals................... 113

9.1	The Complex Relationship Between Stress Sensitivity and the Transmission Mode and Diversity of Symbionts	115
9.2	Do Bacteria Cause Coral Bleaching?........................	116
9.3	Coral Disease and the Significance of Opportunistic Pathogens...	116
9.4	Changes in Coral-Associated Microbial Consortia Under Stress..	117
9.5	*Symbiodinium* as a Recent Intruder on Preexisting Coral–Bacterial Mutualisms	118
9.6	Coda: Are Coral Reefs Doomed?	119
	9.6.1 The Geological Perspective: The Persistence of Coral Reefs	120
	9.6.2 Impacts of Ocean Acidification on Corals...............	122
	9.6.3 What About the Direct Impact of Thermal Stress or Elevated CO_2 on Corals?...........................	123

		9.6.4	Can Corals Evolve Fast Enough to Keep Pace with the Rate of Climate Change?	123
	Conclusion			124
	References			124
10	**The Hidden Impact of Viruses**			**127**
	10.1	Beneficial Viruses		128
	10.2	Viral Communities in *Hydra* Are Species Specific and Sensitive to Stress		129
	10.3	Bacteriophage Therapy in Corals?		131
	Conclusion			132
	References			132
11	**Seeking a Holistic View of Early Emerging Metazoans: The Power of Modularity**			**135**
	11.1	Animals Are Mobile Ecosystems Carrying a Myriad of Microbes with Them		135
	11.2	The Power of Modularity		137
	References			138
Further Reading				**139**
Index				**153**

Introduction: The Holobiont Imperative

This book is being written at a time when fundamental shifts in thinking are occurring in the life sciences, but when the metaphorical ground has not yet settled under our feet. There are no germ-free animals in nature. Epithelia in contact with the environment are colonized by microbial communities, and all multicellular organisms must be considered an association of the macroscopic host in synergistic interdependence with bacteria, archaea, fungi, and numerous other microbial and eukaryotic species. We refer to these associations that can be analyzed, measured, and sequenced, as "holobionts" or "metaorganisms" (Fig. 1.1).

Half a century ago, Lynn Margulis (1993) popularized the idea that symbiosis has been an important factor in evolution, but much of the immediate interest was on the most obvious and significant eukaryote–eukaryote symbioses such as corals and giant clams, the only symbioses involving prokaryotes to receive significant attention being lichens and rhizobia. By contrast, there is now a growing realization of the importance and ubiquity of associations involving prokaryotes and archaea to every aspect of animal life—bacteria not only enable animals to metabolize otherwise indigestible polysaccharides such as lignin and cellulose, but also shape animal development and behavior.

This scenario is also playing out within the field of "traditional" symbioses, so that whereas 20 years ago, the coral symbiosis was viewed simply as a cnidarian–dinoflagellate association, current thinking has the coral "holobiont" beside the photosynthetic algae *Symbiodinium* also including bacteria and also viruses.

Interactions between the members of the holobiont, i.e., bacteria, eukaryotic symbionts, and host cells, have probably been critical to enabling the key transitions in animal evolution. However, the reciprocal is also true—animals have dramatically transformed the physical environment that is available for bacterial colonization. Animals also provide niches that simply do not exist elsewhere—for example, the rumen and the vertebrate gut, the light organ of the bobtail squid, or the intracellular environment of an ascidian. Animals also exercise enormous selective forces on bacterial populations—think only of the spread of multidrug-resistant (MDR)

© Springer-Verlag Wien 2016
T.C.G. Bosch, D.J. Miller, *The Holobiont Imperative: Perspectives from Early Emerging Animals*, DOI 10.1007/978-3-7091-1896-2_1

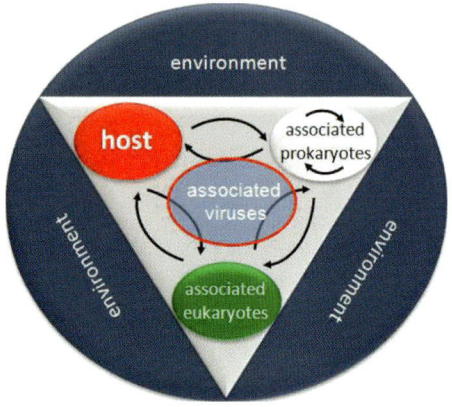

Fig. 1.1 Any multicellular organism must be considered a holobiont or metaorganism, a complex community of many species which have been, and are being, evolved

strains of *Staphylococcus aureus*, or the evolution of bacteria capable of degrading completely novel chlorinated hydrocarbons.

Since their Precambrian origins, the Metazoa have transformed the physical environment, but always in collaboration with bacteria. Along the way, some animals have also formed close relationships with other eukaryotes, but these macrosymbiotic bonds have been forged in the context of preexisting host–bacteria interactions. As discussed in Chap. 5, the partnerships that animals have forged with bacteria have been powerful agents of change. This becomes particularly apparent by the transformation of vegetation as a consequence of the evolution of the ruminants.

The increasing realization that animals cannot be considered in isolation but only as a partnership of animals and symbionts has lead to two important realizations. First, it is becoming increasingly clear that to understand the evolution and biology of a given species, we cannot study the species in isolation. Second, the health of animals, including humans, appears to be fundamentally multiorganismal. Any disturbance within the complex community has drastic consequences for the well-being of the members.

1.1 Of Complex Diseases and Animals as Complex Systems: Why Bacteria Matter

The last 50 years have seen fantastic progress in combating and eradicating terrible diseases. Deaths from infectious diseases have declined markedly in the last 50 years. In 2002, Jean-Francois Bach published a study in the New England Journal of Medicine showing an inverse relationship between the prevalence of infectious diseases (decreasing) and the prevalence of immune disorders (increasing) (Fig. 1.2a). The development of antibiotics and other antimicrobial medicines together with strategic vaccination campaigns has virtually eliminated diseases that previously were common in the United States and Europe, including diphtheria, tetanus, poliomyelitis, smallpox, measles, mumps, rubella, and *Haemophilus*

1.1 Of Complex Diseases and Animals as Complex Systems: Why Bacteria Matter

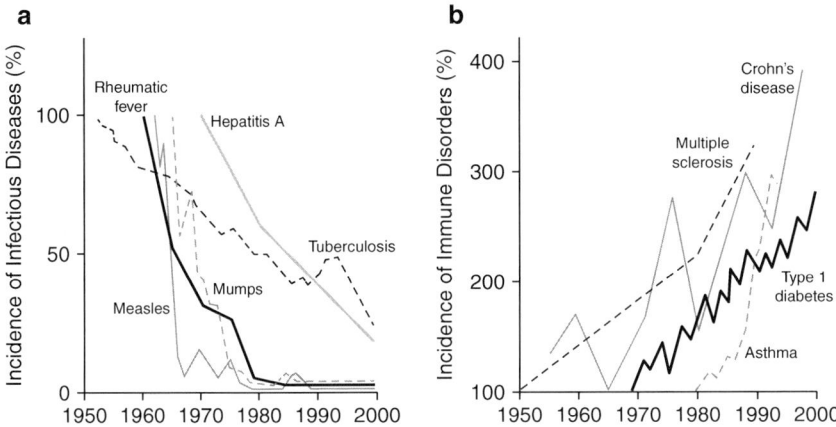

Fig. 1.2 Inverse relationship between the incidence of infectious disease (*left*) and immune disease (*right*) from 1950 to 2000 (Bach 2002)

influenzae type b meningitis. As reported by the National Center for Infectious Diseases, CDC, of the United States in 1999, this decline contributed to a sharp drop in infant and child mortality and also to a significant increase in life expectancy.

However, this success story has a second face (Fig. 1.2b). As pointed out first by Jean-François Bach (2002) from the INSERM Research Institute and Hôpital Necker in Paris, epidemiologic data provide strong evidence of a steady rise in the incidence of allergic and autoimmune diseases in developed countries over the last 50 years. The incidence of many diseases of these two general types has increased: asthma, rhinitis, and atopic dermatitis, representing allergic diseases, multiple sclerosis, and insulin-dependent diabetes mellitus (type 1 diabetes)—particularly in young children—and Crohn's disease, representing autoimmune diseases. The prevalence of asthma, hay fever, and atopic dermatitis doubled in Swedish schoolchildren between 1979 and 1991, and in Lower Saxony, Germany, the incidence of multiple sclerosis also doubled from 1969 to 1986. The incidence of Crohn's disease more than tripled in northern Europe from the 1950s to the 1990s. The incidence of these disorders apparently began to increase in the 1950s and continues to do so today, although the incidence of some of these diseases may have plateaued.

Thus, success in reducing morbidity and mortality from infectious diseases during the first three quarters of the twentieth century, however, came at a cost and was accompanied by the appearance of diseases which were unknown before in humans and animals. Allergies used to be rare condition, but now as many as 1 in 50 persons has the condition. Although most cases are mild, and overdiagnosis is likely, allergic reactions can be severe, sometimes leading to sudden death. The prevalence of both hay fever and eczema has been rising dramatically in recent years, paralleling the increase in asthma and type 1 diabetes. Another condition to consider is what is called inflammatory bowel disease (IBD), a group of chronic, relapsing disorders of the intestine. IBD manifests in two main types, ulcerative colitis and Crohn's disease, which partially overlap but have different pathology.

The etiology of IBD is complex, multifactorial, and incompletely understood. Throughout the past century, many theories have proposed and/or implicated the role of different bacteria. In particular, microbial dysbiosis has been hypothesized as a key player in disease development (Dasgupta and Kasper 2013).

Studies that have examined the role of altered microbiota in IBD demonstrate reduced gut microbiome richness and biodiversity, such as a decrease in *Faecalibacteria* with *Faecalibacterium prausnitzii* in mucosa-associated microbiota or feces. A definite causal relationship between bacteria and the pathogenesis of IBD has not yet been identified. Two recent observations and developments point to causal relationships rather than simple associations between the microbiome and IBD. First, nearly every mouse model of the disease requires the presence of microbes for colitis to develop. And second, in humans, fecal microbiota transplantation (FMT) turns out as a safe, but variably efficacious novel treatment option for inflammatory bowel diseases (IBD) (Colman and Rubin 2014). Although we may not know yet which is cause and which is effect—are the microbes causing the disease or not and is there a proof of Koch's postulates for IBD or is it elusive—these studies show that microbes are somehow involved. And then there is autism. When the disorder was first described in 1943 by Dr. Kanner, it was uncommon. Today, about one in 88 children has autism or autism spectrum disorder (ASD). Although overdiagnosis certainly contributes to the rise in cases, it is not enough to explain the enormous increase. Multiple theories try to explain the increase in autism cases, including toxins in water, food, and air and exposures to chemicals and pesticides during pregnancy. But no one knows for sure. Correlation does not equal causation. Evidence, however, is mounting that intestinal microbes exacerbate or perhaps even cause some of autism's symptoms. Recent observations in animal models show not only that gut microbes are involved in brain development but also that autism-like syndromes are not developing in germ-free animals, and that syndromes can be cured by addition of certain bacterial compounds. Again, it is not clear whether these microbial differences drive the development of the condition or are instead a consequence of it. A study published in December 2013 in cell (Hsiao et al. 2013) supports the former idea. When researchers at the California Institute of Technology incited autism-like symptoms in mice using an established paradigm that involved infecting their mothers with a viruslike molecule during pregnancy, they found that after birth, the mice had altered gut bacteria compared with healthy mice. By treating the sick rodents with a health-promoting bacterium called *Bacteroides fragilis*, the researchers were able to attenuate some, but not all, of their behavioral symptoms. The treated mice had less anxious and stereotyped behaviors and became more vocally communicative. Taken together, the composition of the gut microbes and their metabolic activity seems to be an important factor to keep in mind when attempting to understand why the incidence of autism is increasing so dramatically in the last few years.

It seems, therefore, that we have traded the eradication of infectious diseases with the appearances of immune deficiencies, allergies, asthma, and inflammatory bowel disease (Fig. 1.2). The reason for that is not fully understood. However, since any animal and any human individual appears to be in very close interaction with a stable microbiota, there must be an enormous crosstalk going on between the

1.1 Of Complex Diseases and Animals as Complex Systems: Why Bacteria Matter   5

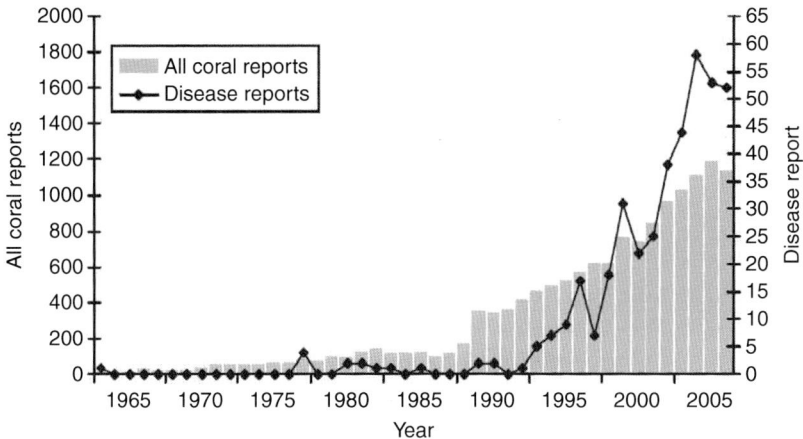

Fig. 1.3 Number of coral disease reports (excluding noninfectious bleaching) compared with all other coral reports over time (Source: Sokolov (2009))

symbionts and the host cells. Any disturbance of this crosstalk may result in severe disturbances. This phenomenon is not limited to the human population, but a worldwide phenomenon and a characteristic feature which can be traced back even to one of the most simple forms of multicellular animal life, the coral polyps.

Susanne Sokolow, a researcher working at the University of California, Santa Barbara, with interest in infectious disease ecology in marine and aquatic ecosystems, compiled a list of all articles about coral disease, published up to December 2008, since the first coral disease report in 1965 (excluding those pertaining only to stress-induced bleaching) (Sokolov 2009). This list was compared to all reports from the same time period retrieved in the ISI Web of Science using the search word "coral." Both disease and non-disease reports exponentially increased over the observation period. As shown in Fig. 1.3, Sokolow's findings indicate that coral diseases (not just bleaching) are emerging and also that coral disease research is rapidly expanding. Thus, these simple creatures also suffer from complex diseases that have increased in the last 50 years (Fig. 1.3).

Most dramatically and visible to any tourist snorkeling in a reef is a disease termed "coral bleaching" (Fig. 1.4). Coral bleaching is the loss of intracellular endosymbionts (*Symbiodinium*, also known as zooxanthellae) through either expulsion or loss of algal pigmentation. Bleaching occurs when the crosstalk between the symbionts and the coral cells is disturbed and the conditions necessary to sustain the coral's zooxanthellae cannot be maintained.

The fact that we and obviously all multicellular organisms coexist with bacteria (for reference see, e.g. Knowlton and Rohwer (2003)) tells us that our microbial "companions" may be there for a reason. Everything that changes the symbiotic partners appears to have a potential cost to us. That is obviously how we have evolved. This raises a profound set of questions. Why do we tolerate them? How do we achieve a stable partnership with our microbes? And how do the microbes manage to live with us for such a long time?

Fig. 1.4 Bleached *Acropora* coral (foreground) and normal colony (background), Keppel Islands, Great Barrier Reef (Taken from en.wikipedia.org)

The seemingly stable partnership bears—as all partnerships—a fundamental tension between conflict and cooperation. In fact, the realization of the ubiquity of mutualisms involving bacteria and animals is one aspect of the "shifting ground" referred to above, but another is the realization that strict classical definitions of "symbiosis"—involving mutual benefits—fail to capture almost all real-world interactions. In many cases, the assumption of mutual benefits has not been rigorously tested, and classical "symbioses," such as that between the reef-building coral and dinoflagellates of the genus *Symbiodinium*, are more appropriately viewed as a spectrum of interactions ranging from symbiotic to outright parasitic in nature. For understanding the behavior of the members of this apparently coevolved association, it turns out to be helpful to remember the game theory by the great economist and mathematician John Nash. His theory sheds light on the phenomenon of cooperation, on how coevolved systems appear to select for individuals who largely play by the rule. And it basically makes the point that cooperation is a strenuous business. If you mess it up or cheat, your outcome is worse that if you played fair. Can that be applied to microbial communities as well? Microbial communities display a variety of selection scenarios that reveal constant selection pressures, but also often comprise a degree of complexity that can only be captured by frequency-and density-dependent selection pressures (Fraune et al. 2014). Frequency- and density-dependent selection pressures give rise to the study of competition and conflict and cooperation and coexistence in interacting bacterial communities. And such complex phenomena in ecological time scales are aptly described by evolutionary games (Li et al. 2015).

1.2 The Complexity of Coevolved Animal Communities Was Discovered in 1877 in Kiel, Germany

In 1877, in one of the first studies to be conducted in the emerging science of ecology, a professor newly recruited to the University of Kiel in the North of Germany, Karl August Möbius (Fig. 1.5), was seeking to determine why the oyster beds of

1.2 The Complexity of Coevolved Animal Communities Was Discovered in 1877

Fig. 1.5 Karl August Möbius (1825–1908) was an extraordinary zoologist of the second half of the nineteenth century. His research on corals and foraminiferans (i.e., protozoans of the rhizopodan order Foraminiferida) led to the discovery of symbiosis in marine invertebrates. Karl August Möbius's conception and studies of "biocenosis," or living community, and thus his contributions in founding both, ecology and marine biology, are gaining recognition only recently (Copyright: Kiel University)

Cancale, Marennes, and Arcachon were becoming exhausted, while the oyster beds in the British river estuaries and the Schleswig-Holstein oyster beds were very rich. In his 1877 work "Die Auster und die Austernwirtschaft" (The Oyster and Oyster Farming), he gave a detailed description of the interaction between oysters and other plants and animals in an oyster bank. Möbius had recognized the interdependence of the oysters and surrounding life forms and coined the term "biocenosis" or "living community" "for a group of mutually dependent species and individuals, the variety and number of which are determined by the average external living conditions and sustained in an appropriate area by means of reproduction."

Thus, Möbius related the flourishing (as well as the wasting away) of oysters in a given habitat to the other species present, rather than to the oysters in the beds themselves. This was a fundamental discovery. Möbius was the first to recognize that an ecological system must be taken as a whole. His biocenosis theory established itself as the basis of general ecology.

Similar observations were made about a hundred years later in Jamaica's coral reefs (Kaufman 1983). There it was the decline of the entire population of long-spined sea urchins which in 1983 caught the attention of reef ecologists and environmentalist (Knowlton 2001). An unidentified pathogen killed most of the sea urchins and resulted in an enormous overgrowth of algae, eventually covering more than 90 % of Jamaica's reef surface. This had catastrophic consequences such as killing the corals underneath. Similar catastrophes happened in many other places of the world. In all instances, a carefully balanced partnership between several members of a coevolved community—biocenosis—was all over suddenly getting out of control. Consequently, medical doctors, by analogy with "biocenosis," began to conceptualize the understanding of a disease as a complex dynamic phenomenon with the word "pathocenosis" (Grmek 1969).

Today, we are convinced that not only ecological systems but also complex "environmental" diseases can only be understood if the relationships between the interacting infectious agents present at a given time in a given territory are recognized. *"We like to see the world as consisting of separate parts that can be studied in an isolated, linear way, one piece at a time. These pieces then can be summed independently to make the whole,"* says George Sugihara, a mathematician and theoretical ecologist at the Scripps Institution of Oceanography in San Diego (May et al. 2008). *"The trouble and real danger is that we persist with these linear tools and models even when systems that interest us are complex and nonlinear."*

1.3 Looking for a Term for the Functional Entity Formed by a Host and Its Associated Microbial Symbionts

Today, we realize that any multicellular organism must be considered an association comprised of the macroscopic host and synergistic interdependence with bacteria, archaea, fungi, and numerous other microbial and eukaryotic species. We specifically refer to these associations as "holobionts" (or sometimes "metaorganisms"), because this collective term defines a superordinate entity that is applicable to all kinds of interdependent associations. The metaorganism concept considers the dynamic communities of bacteria on epithelial surfaces as an integral part of the functionality of the respective organism itself. "Holobiont" and its synonym "metaorganism" refer to physical associations of organisms that can be analyzed, measured, and sequenced. The term "metaorganism" was first used by Graham Bell (1998) to refer to organisms which are between two levels of organization. Recently, the term is increasingly used to refer to the totality of any multicellular organism derived from millennia of coevolution with microbiota (Biagi et al. 2012). Even humans have been reviewed as "metaorganisms" as a result of a close symbiotic relationship with the intestinal microbiota (Turnbaugh et al. 2007). Scientists around Jeffrey Gordon, Nancy Knowlton, David A. Relman, Forest Rohwer, and Merry Youle have used the April 2013 issue of the American Society of Microbiology (ASM) "Microbe magazine" for proposing to use for the functional entity formed by a macrobe and its associated symbiotic microbes and viruses the term "holobiont." Both terms, "metaorganism" and its synonym "holobiont" refer to an association of organisms which occupies an ecological niche, adapts, and as discussed in Chap. 4 of this book may even be the organizational level at which natural selection acts.

Here, we enter controversial territory. Can adaptation occur by swapping microbial constituents or by reshuffling the relative proportions of current bionts? Will natural selection then favor the holobionts with constituents that confer increased fitness under the new conditions? Can, therefore, a holobiont employ strategies unavailable to any one species alone? Should we consider ourselves—as suggested by Jeffrey Gordon and colleagues—*"with humility and with deep appreciation as but one biont within the human holobiont?"*

In this book, we limit our scope to symbioses (*sensu lato*) involving animals and, although we may throw in a few colorful examples from the discipline, we do not attempt to cover in depth the medical literature—this is enormous, we are not experts, and a lot has already been written on the subject. This is a book about the complexity of early emerging metazoans. When searching for general concepts, simple animal models such as cnidarians may help to study mechanisms and identify key players or mediators even when they only partly reflect the human situation.

We wrote this book for the nonspecialist. Each chapter is an independent contribution, mostly summarizing what we have investigated in our laboratories in the last few years. We will focus on the discovery of microbes as partners in early animal evolution and development; we then will consider the mechanisms controlling and regulating this partnership; and finally, we will present facts and ideas on the selective advantage of such host–microbe interactions and the expanded capabilities that these associations confer so that they have been maintained over millions of years. The ten chapters should provide the material for the conclusions of the last essay on Chap. 11 on "the power of modularity." If the reader finds the inclination and leisure to read through this final one, we have reached our goal. The scope of this book is best described as an attempt to understand animal evolution in terms of symbiotic interactions and in light of the realization that we animals are intruders that have evolved within a microbial world. In order to do that, we first must understand where we come from. So that is where we will turn first.

References

Bach JF (2002) The effect of infections on susceptibility to autoimmune and allergic diseases'. New Engl J Med 347(2002):911–920

Bell G (1998) Model metaorganism. A book review. Science 282(5387):248

Biagi E, Candela M, Fairweather-Tait S, Franceschi C, Brigidi P (2012) Ageing of the human metaorganism: the microbial counterpart. Age (Dordr) 34(1):247–267

Colman RJ, Rubin DT, 12 (2014) Fecal microbiota transplantation as therapy for inflammatory bowel disease: a systematic review and meta-analysis. J Crohn's Colitis 8:1569–1581

Dasgupta S, Kasper DL (2013) Relevance of commensal microbiota in the treatment and prevention of inflammatory bowel disease. Inflamm Bowel Dis 19:2478–2489

Fraune S, Anton-Erxleben F, Augustin R, Franzenburg S, Knop M, Schröder K, Willoweit-Ohl D, Bosch TC (2014) Bacteria-bacteria interactions within the microbiota of the ancestral metazoan Hydra contribute to fungal resistance. ISME J 9(7):1543–1556. doi:10.1038/ismej.2014.239

Grmek MD (1969) Préliminaires d'une étude historique des maladies. Ann Econ Soc Civilisat 24:1437–1483

Gordon J, Knowlton N, Relman DA, Rohwer F, Youle M (2013) Superorganisms and holobionts. Microbe magazine. April 2013 issue

Hsiao EY, McBride SW, Hsien S, Sharon G, Hyde ER, McCue T, Codelli JA, Chow J, Reisman SE, Petrosino JF, Patterson PH, Mazmanian SK (2013) Microbiota modulate behavioral and physiological abnormalities associated with neurodevelopmental disorders. Cell 155(7):1451–1463

Kaufman LS (1983) Effects of Hurricane Allen on reef fish assemblages near Discovery Bay, Jamaica. Coral Reefs 2:1–5

Knowlton N (2001) The future of coral reefs. Proc Natl Acad Sci U S A 98(10):5419–5425

Knowlton N, Rohwer F (2003) Multispecies microbial mutualisms on coral reefs: the host as a habitat. Am Nat 162:S51–S62

Li XY, Pietschke C, Fraune S, Altrock PM, Bosch TC, Traulsen A (2015) Which games are growing bacterial populations playing? J R Soc Interface. 12(108):20150121.

Margulis L (1993) Symbiosis in cell evolution, 2nd edn. W. H. Freeman & Co., New York

May RM, Levin SA, Sugihara G (2008) Complex systems: Ecology for bankers. Nature 451:893–895

Möbius KA (1877) Die Auster und die Austernwirtschaft. Wiegandt, Hempel & Parey, Berlin

Sokolov S (2009) Effects of a changing climate on the dynamics of coral infectious disease: a review of the evidence. Dis Aquat Organ 87(1–2):5–18

Turnbaugh PJ, Ley RE, Hamady M, Fraser-Liggett CM, Knight R, Gordon JI (2007) The human microbiome project. Nature 449:804–810

Major Events in the Evolution of Planet Earth: Some Origin Stories

2.1 Microbes Were First: Bacteria Have Existed from Very Early in the History of Life on Earth

With billions of years of evolution before the appearance of animals, prokaryotes shaped and continue to shape both the Earth's biogeochemical landscape and the setting for animal existence (Fig. 2.1) (Knoll 2003).

Bacteria inhabit every environment on the planet. Bacteria fossils discovered in rocks date from at least the Devonian Period, and there are convincing arguments that bacteria have been present since early Precambrian time, about 3.5 billion years ago (Fig. 2.2).

This long history has resulted in ecological interactions among microbes that are broadly diverse and flexible, features enabled by their rapid generation times and large population sizes, in addition to their proclivity for horizontal gene transfer (HGT).

Since animals diverged from their protistan ancestors some 3 billion years after bacterial life originated and as much as 1 billion years after the first appearance of eukaryotic cells, relationships of animals with bacteria were likely already operating when animals first appeared near the end of the Proterozoic Eon. Animal evolution, therefore, is intimately linked to the presence of microbes.

2.2 Life Did Not Take Over the Globe by Combat, But by Networking

Microbes can be critical determinants of animal population and community structures. Larvae of different marine invertebrate species, for example, are known to respond to either dissolved or surface-bound cues to settlement and metamorphosis. Researchers in the Hadfield laboratory in the Kewalo Marine Lab, a unit of the University of Hawaii, have shown that many of these settlement cues are produced by bacteria (Hadfield 2011). And on the other hand, although most of the microbial

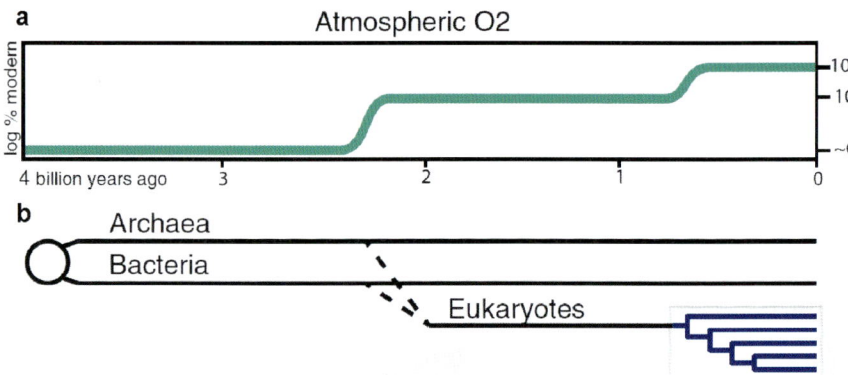

Fig. 2.1 The evolution of plants and animals in Earth's history has occurred as a patina upon a microbe-dominated landscape. (**a**) Upper atmospheric oxygen concentration, as a percent of current levels, plotted against geological time. (**b**) Phylogenetic history of life on Earth, scaled to match the oxygen timeline. Note that the origin of the eukaryotes and the subsequent diversification of animals both correspond to periods of increasing atmospheric oxygen (Source: McFall-Ngai et al. 2013)

Fig. 2.2 Microscopic view of cyanobacteria, a life-form that first evolved on our planet 3.5 billion years ago became widespread and quite abundant during the Proterozoic (Image Source: Bruce Foulke, U.S. Office of Naval Research)

world operates independently of animals, animals can profoundly impact co-occurring microbes in communities and ecosystems. Examples include the introduction of European earthworms to a North American forest which led to massive declines in soil–microbe biomass and respiration. And rats introduced onto small Pacific islands decimated seabird populations, resulting in decreased guano input, which altered decomposition and nutrient cycling by soil microbes and created a dramatic reduction in ecosystem productivity (Fukami et al. 2006).

Historically, bacteria are seen as pathogens. Starting with Louis Pasteur's discovery of the link between germs and disease, microorganisms were of interest mainly because they were considered to be causes of infectious diseases. In the last 100 years, numerous microorganisms were identified as the causative agents of important human diseases, and bacteriologists, microbiologists, and immunologists are continuing to focus on bacteria as pathogens. This approach has led to enormous insights in the battle between the invading harmful microbes and the host, as well as enabled the development of efficient strategies to fight infections. However, most bacteria are not harmful to plants or animals, and many of them are beneficial, playing a key ecological role. Louise Taylor and colleagues from the Centre for Tropical Veterinary Medicine at the University of Edinburgh have used the published literature to compile a list of organisms known to be pathogenic to humans, together with the available information on whether they are zoonotic, whether they are regarded as emerging, and on their transmission routes and epidemiologies. Their estimate indicates that only 200 of the millions of bacteria that interact with humans are pathogenic (Taylor et al. 2001).

In fact, eukaryotic cells per se are the descendents of separate prokaryotic cells that joined together in an endosymbiotic event with mitochondria being the direct descendents of a free-living bacterium that was engulfed by another cell (for recent review see Keeling et al. 2015). This became evident in the 1970s when scientists developed new tools and methods for comparing genes from different species. Two teams of microbiologists—one headed by Carl Woese (Woese and Fox 1977) and ther other by W. Ford Doolittle at Dalhousie University in Nova Scotia—studied the genes inside chloroplasts of some species of algae. They found that the chloroplast genes bore little resemblance to the genes in the algae's nuclei. Chloroplast DNA, it turns out, was cyanobacterial DNA. The DNA in mitochondria, meanwhile, resembles that within a group of bacteria that includes the type of bacteria that causes typhus.

The endosymbiotic theory was advanced and substantiated with microbiological evidence by Lynn Margulis in a 1967 paper. According to Margulis and Dorion (2001), "Life did not take over the globe by combat, but by networking" (i.e., by cooperation). It has become clear that symbiotic events have had a profound impact on the organization and complexity of many forms of life. Algae have swallowed up bacterial partners and have themselves been included within other single cells. Nucleated cells are more like tightly knit communities than single individuals (Doolittle et al. 2013; Doolittle and Zhaxybayeva 2009). Evolution is more flexible than was once believed.

2.3   The Transformation of the Biosphere at the Ediacaran–Cambrian Boundary

The early Earth was a hostile place, but for most of its 4.6 billion years of its history, life has been present (Fig. 2.3). Prokaryotic life dates back around 4 billion years, its emergence following closely after the end of the period of heaviest bombardments. While the earliest life was unicellular, experiments in multicellularity occurred very early in the history of life on Earth. Some of early Archaean (3.5 BYA) fossils provide clear evidence for early experiments in multicellularity, and by the late Proterozoic (850 MYA), a diverse range of multicellular eukaryotes was present.

Animal life came much later; in fact, fossils from the mid-late Ediacaran (around 580 MYA) represent the earliest evidence for macroscopic multicellular life of any kind and the spectacular radiation of extant animal phyla, known as the Cambrian explosion, at 543 MYA.

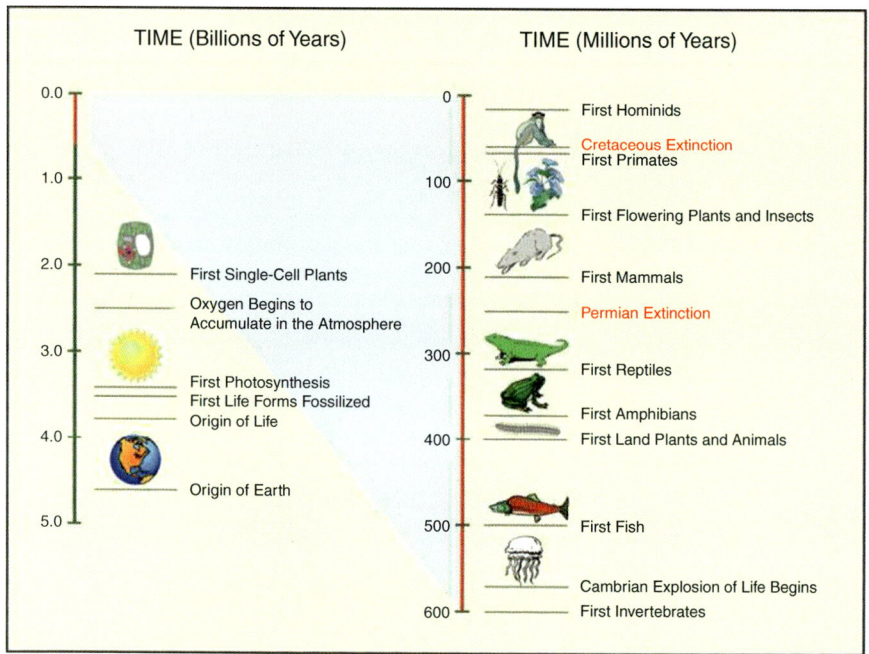

Fig. 2.3 Important events in the evolution of life. Dates for many of the events shown are based on fossil evidence (Image Copyright: Michael Pidwirny) (Taken from The Encyclopedia of Earth, http://www.eoearth.org/view/article/155427/)

Why did it take so long for animals to arrive? The answer that this question usually invokes is that levels of oxygen in solution in ancient oceans, which had been increasing gradually, suddenly breached a threshold that could sustain the greater demands (motility, etc.) of animal life. Since the great oxygenation event, the activities of cyanobacteria had brought about steady increases in atmospheric O_2, so that by the late Vendian and early Cambrian, atmospheric O_2 was probably near present-day levels. The Ediacaran animals were restricted to those patches of shallow water that were relatively well oxygenated; the deeper oceans were in transition from being predominantly anoxic and sulfide-rich to being anoxic but iron-rich and possibly also rich in dissolved organic carbon, but still hostile to all but microbial life.

Reaching a critical atmospheric oxygen level is often nominated as the likely trigger for the emergence of complex multicellular animals, because metazoans have a relatively high demand for oxygen; higher O_2 levels may also have liberated the ancient ecosystem from nitrogen starvation imposed by the sulfide-rich and anoxic state of the oceans.

The idea of a critical level of oxygen triggering animal evolution is, however, not unanimously subscribed to. An attractive alternative idea is that early animals themselves were the agents of change—that by greatly increasing ventilation of the oceans, early animals actually caused most of the observed geochemical perturbations observed during the Neoproterozoic–Paleozoic transition, rather than the geochemical events enabling/"causing" animal evolution. Zooplankton grazing may have selected for large phytoplankton that were export prone (i.e., upon demise, prone to sinking to the benthos and thus exporting production from the photic zone), effectively transforming what was a cyanobacteria-dominated and anoxic water column that was probably also turbid and stratified into a mixed and aerated system dominated by eukaryotic algae. By causing mixing of the oceans, animals started to transform the environment and thus not only to alter the distribution of microbes but also to open new niches for microbial colonization. Animals, therefore, have thus been major agents of change since their early origins.

2.4 Our Bacterial Ancestry Is Reflected in Our Genomic Signature

Genome sequences provide a unique window into a long history of evolution and dependence on a given habitat. Until quite recently, the consensus view was that animals possessed many genes and pathways not present in other organisms and that an overall increase in genetic complexity (i.e., kinds and numbers of genes) was likely to correlate with increasing morphological sophistication. This view of the ascent of man is almost Copernican in its arrogance, as become clear in the early 2000s with the availability of large molecular datasets for morphologically simple animals.

One of the most intriguing findings to emerge from the sequencing projects in many different animal taxa is that the gene sets of the common metazoan ancestor are surprisingly rich and complex. This flatly contradicts traditional expectations—modern cnidarians are "primitive" animals, with little obvious morphological differentiation, and this is often also assumed to have been the case with Urbilateria. The assumption has been that simple morphology equates to a simple gene set; fewer genes should be required to build a sea anemone than a fly, but this seems not to be true. Sequencing data from cnidarians in particular, but also from sponges, revealed the presence of many genes previously only known from vertebrates and assumed to have evolved in the chordate lineage on the basis of absence from the genomes of the only other metazoans for which data were available at that time, the fruit fly *Drosophila* and the worm *Caenorhabditis*. The cnidarian data in particular have led to the broad (but not universal) acceptance of the idea that genetic complexity came very early in animal evolution, but that widespread gene losses and sequence divergence have occurred and affected some lineages more than others.

While this discovery rapidly remodeled how biologists view the diversity of life, it is only in the last decade that advances in molecular technology have created a seismic shift in biological worldview. Enabled by new genomic approaches, we can now specifically identify microbes and define their activities within communities. The lesson from the comparative genomics of higher animals is that their genomic "dictionaries" share common and deep evolutionary ancestry. This major discovery became possible when a minimum number of genomes had been fully sequenced, the analyses of which are totally reshaping the way biologists view relationships in

Fig. 2.4 Our bacterial ancestry is reflected in our genomic signature. Image illustrates the relative percentage of human genes that arose at a series of stages in biological evolution. By intertwining of animal and bacterial genomes, a host can expand its metabolic potential, extending both its ecological versatility and responsiveness to environmental change (Source: McFall-Ngai et al. 2013)

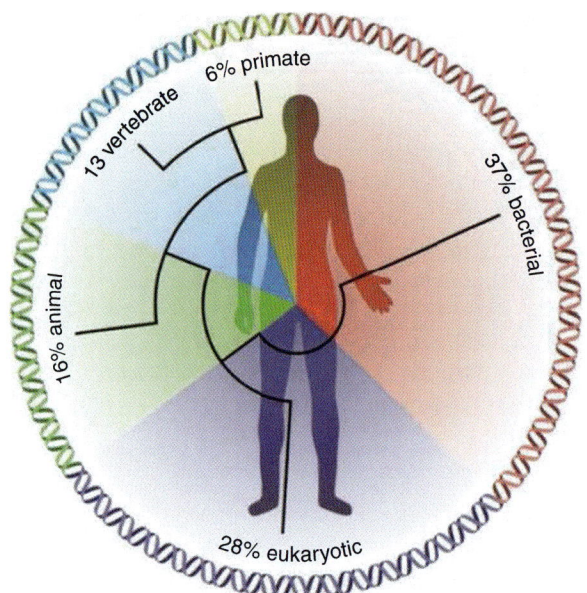

the biosphere. Examination of these genomic data reveals that essential all life forms share ~1/3 of their genes, including those encoding central metabolic pathways. Many animal genes are either direct descendants of microbial genes or the result of horizontal gene transfer (HGT). For example, 60 % of the ~23,000 human genes trace back to at least the origin eukaryotes (Fig. 2.4). The products encoded by these shared genes may provide the foundation for much of the signaling and communication among extant taxa of these two divergent groups.

2.5 Genomes of Early Emerging Metazoans, Similar to Humans, Contain a Considerable Fraction of Genes Encoding Proteins of Bacterial Origin

Marine sponges often contain dense and diverse microbial communities, which can constitute up to 35 % of the sponge biomass. Interestingly, metagenomic analysis of sequence reads of the genome of the sponge *Amphimedon queenslandica* (Fig. 2.5) is consistent with existence of a dominant proteobacterial symbiont. From a randomly chosen representative subset of unassembled, filtered reads (120,000 out of 217,873 total), 7720 (6.4 %) were putatively assigned to the bacterial domain of life, and a very small number were assigned to archaea (161, 0.1 %) (Srivastava et al. 2010). The majority of reads map to alpha and gammaproteobacteria. The fraction of reads assigned to alphaproteobacteria exceeds the fraction of all sequenced genomes that belong to alphaproteobacteria. Likewise, we find an excess of reads assigned to Planctomycetes relative to the number of genomes sequenced; however, the total number of putatively bacterial reads assigned to Planctomycetes

2.5 Genomes of Early Emerging Metazoans, Similar to Humans

Fig. 2.5 The sea sponge *Amphimedon queenslandica* is the first sponge to have its genome sequenced. This demosponge is found in shallow waters of the Great Barrier Reef and is common around Heron Island (Source: degnanlabs.info)

is small (2.8 %). Thus, metagenomic analysis of sequence reads is consistent with existence of a dominant proteobacterial symbiont.

Evidence for host–microbe interactions has also been published in the course of the draft assembly of the dinoflagellate *Symbiodinium minutum* genome (Shoguchi et al. 2013). Photosynthetic symbionts such as *Symbiodinium* are essential to reef building. The estimated nuclear genome of this species is 1500 Mbp and contains 42,000 protein-coding genes. Similar to observations in *Hydra* and *Nematostella*, the *Symbiodinium* genome provided evidence for the close association with alphaproteobacterium. The largest scaffold in the assembly appears to form a circular DNA of 3.8 Mbp. The GC content of the scaffold was 53 %, compared to 44 % in *S. minutum*. In addition, no expressed sequence tag (EST) data were mapped onto the putative bacterial genome fragment, indicating that the source of this DNA was quite different from *Symbiodinium*. Phylogenetic analysis with the contaminant 16S ribosomal RNA sequence indicated that this organism shows the closest match to *Parvibaculum lavamentivorans* DS-1. *P. lavamentivorans* DS-1 T is the type of species of the novel genus *Parvibaculum* in the novel family Rhodobiaceae (formerly Phyllobacteriaceae) of the order Rhizobiales of alphaproteobacteria. Scanning-electron microscopy shows the presence of bacteria on the surfaces of *Symbiodinium* cells and therefore supports the view that even a protest in fact may be closely associated with bacteria throughout most or all of the host life cycle and thereby should also be considered a holobiont.

The sequence of the *Hydra magnipapillata* genome (Chapman et al. 2010) opened a first window into the *Hydra* holobiont since the genome assembly yielded eight large putative bacterial scaffolds as evidenced by high G+C content, no high-copy repeat sequences typical of *Hydra* scaffolds, and closely spaced single-exon open reading frames with best hits to bacterial sequences. These scaffolds span a total of 4 Mb encoding 3782 single-exon genes and represent an estimated 98 % of the bacterial chromosome. Phylogenetic analysis of 16S rRNA and conserved clusters of orthologous groups of proteins indicate that this bacterium is a novel *Curvibacter* species belonging to the family Comamonadaceae (order Burkholderiales). About 60 % of annotated *Curvibacter sp.* genes have an ortholog in another species of Comamonadaceae. Notably, the *Curvibacter sp.* genome encodes nine different ABC sugar transporters, compared to only one or two in other species of Comamonadaceae, possibly reflecting an adaptation to life in association with *Hydra*.

Intriguingly, similar findings were reported after looking more closely into the *Nematostella* genome. In 2007, a shotgun genome sequence of *Nematostella vectensis* was obtained, assembled into long scaffolds, and annotated, and a remarkable resemblance of the gene repertoire in the species to the gene sets in more complex metazoa was emphasized, including a higher sequence similarity, a larger content of shared genes, and a higher degree of synteny between *N. vectensis* and vertebrates than between vertebrates and familiar invertebrate model organisms, such as the fruit fly and a soil nematode. Detailed, case-by-case analysis of regulatory and signaling pathways shared by *N. vectensis* and higher animals has confirmed a Premetazoan origin for many of these pathways, either in substantially complete form or in simpler versions that have been elaborated later in metazoan evolution by molecular "tinkering."

Irena I. Artamonova from the Vavilov Institute of General Genetics in Moscow, Russia, and Arcady R. Mushegian from the National Science Foundation, Arlington, Virginia, USA, have discovered that the *N. vectensis* genome sequence submitted to the sequence databases is very likely to contain many genes of bacterial and bacteriophage origin, derived from prokaryotes closely associated with the sea anemone. The bacterial and phage genes are located in distinct scaffolds, which can be separated from those corresponding to the DNA of the target organism, *N. vectensis*, by several criteria, including the nucleotide composition and the provenance of neighboring genes; the spurious character of introns annotated in these genes; and a rich and apparently randomly drawn repertoire of predicted protein functions, which include several molecular roles considered to be prokaryote specific. The majority of bacterium-like genes from the *N. vectensis* genome project are phylogenetically closer to Bacteroidetes or Proteobacteria homologs than to any eukaryotic homolog. Two distinct bacterial genera are clear leaders in high-sequence similarity of these bacterium-like proteins. They are *Pseudomonas* and *Flavobacterium* (80–85 % and 85–95 % median identity of amino acids to their nearest database homologs, respectively), separated by many hundreds of millions of years of bacterial evolution. Thus, a considerable fraction of *N. vectensis* genes in fact are bacterium-like genes annotated as belonging to *N. vectensis* but encode proteins of bacterial origin. The phylogenetically closest database matches of these bacterium-like sequences originated from several clades of bacteria, but nearly two-third of them belonged to one of two clades: Proteobacteria and Bacteroidetes.

Following the same analytical sequencing approach indicated that the initial set of automatically assigned putative virus proteins consisted of seven bacteriophage proteins and six proteins from eukaryotic viruses.

2.6 The CRISPR/CAS System as Window into Ancient Holobionts

The CRSPR/CAS system is a complex resistance mechanism described in bacteria and is the only adaptive and inheritable prokaryotic immune system. CRISPR stands for "clustered, regularly interspaced, short palindromic repeats" and CAS for "CRISPR-associated genes." In 1987, Ishino and his colleagues discovered an unusual structure of repetitive DNA downstream of the *E. coli inhibitor of apoptosis* (*iap*) gene, consisting of invariant direct repeats and variable spacing elements. In 2002, Jansen and colleagues coined the term CRISPR and reported that CRISPSs colocalized with specific *cas* genes. CRISPS/CAS systems are exclusively found in prokaryotes and are present in approximately half of all bacteria and almost all archaea (Grissa et al. 2007). Analysis of bacterial and archaeal genome sequences and the associated viral and plasmid sequences led to the realization that CRSIPS spacers, which resemble fragments of foreign genetic elements, were derived from invading genomes (Barrangou et al. 2007). This breakthrough, together with the detection of CRISPR locus transcripts with defined length of one or more spacer repeat units, and the predicted nucleic acid-related activities for many of the *cas* genes, led the team of computational biologists around Eugene Koonin at the National Center for Biotechnology Information (NCBI) in Bethesda, to propose that CRIPS/CAS neutralizes invaders via a mechanism reminiscent of RNA interference (Makarova et al. 2011). Soon after, Rodolphe Barrangou, Philippe Horvath, and colleagues provided the first experimental evidence that the CRIPS/CAS system of the lactic acid bacterium *Streptococcus thermophilus* functions as an inheritable, adaptive prokaryotic immune system conferring phage resistance. The bacterial genome integrates a sequence of the viral genome, called a spacer, upon infection; that sequence later serves as a guide for destroying any matching DNA, so that subsequent viral infections are fended off (Horvath and Barrangou 2010).

As pointed out by Artamonova and Mushegian (2013), many of the bacterium-like proteins annotated by the *Nematostella vectensis* genome project have functions that are relevant to the biology of bacteria but have never been reported in eukaryotes. Interestingly, among the examples are proteins with high similarity to Cas1 (one of the CRISPR-associated proteins participating in a prokaryote-specific immunity system), bacterial transcription regulators and bacterial-type protein kinases, the pilin glycosylation protein PglD, OmpA/MotB domain proteins that operate in the outer membrane of gram-negative bacteria, and the aforementioned cyclopeptide biosynthesis enzymes. Even if the genes encoding these proteins were once horizontally transferred to the *N. vectensis* genome, it would be highly unusual for a eukaryote to maintain all of them, in an essentially unmodified form, to perform a set of functions that are without precedent in metazoa.

Artamonova and Mushegian take the finding of CRISPR-associated proteins as computational evidence that argues strongly for the existence of bacteria and viruses closely associated with *N. vectensis*. A substantial, perhaps nearly randomly sampled portion of the *Nematostella* metagenome is already deposited in the databases, apparently having been misannotated as *N. vectensis* genes when in fact this gene complement should be studied further as evidence of the holobiont organization of the sea anemone. Interestingly, the bacterium-like genes revealed by the *N. vectensis* genome project partition from the host genes similarly to the genes of the known bacterial endosymbiont of *Hydra magnipapillata*. Similar to what has been observed previously with *Hydra magnipapillata* bacterium-like genes (Chapman et al. 2010), the genes of the *N. vectensis* bacteria tend to be intronless open reading frames, which are located in separate genomic scaffolds characterized by distinct GC contents.

All these lines of evidence suggest that of the two possible explanations for the presence of these genes in the data banks, i.e., domestication by an invertebrate of horizontally transferred bacterial genes in the evolutionary past and the occurrence of bacteria in the *N. vectensis* planula used for DNA isolation, the latter seems to be more plausible. Far from these genes being biologically irrelevant sample contamination, their presence in the genome database indicates that, similar to the cases of other marine Anthozoa (Williams et al. 2007) and the completely sequenced freshwater hydrozoan *H. magnipapillata*, the body of the starlet sea anemone is in fact a holobiont, i.e., a consortium of a metazoan animal and bacteria closely associated with it throughout most or all of the host life cycle (see also Sect. 6.4).

2.7 Origins of Complexity: What Makes an Animal?

Multicellularity has evolved independently many times and appears to be a trivial undertaking in terms of biology. Under artificial selection in the laboratory, normally unicellular bacteria can achieve a kind of multicellularity by "learning" to adhere to each other. And some bacterial lineages even have achieved a degree of cellular differentiation; for example, when *Nostoc*, a cyanobacterium found in a variety of environmental niches (Fig. 2.6), is grown under N-limitation, nitrogen-fixing heterocysts develop at regular intervals along a chain of "normal" vegetative cells, allowing the oxygen-sensitive nitrogen fixing and oxygenic photosynthesis reactions to be partitioned between functionally differentiated cells. Despite billions of years of opportunity, this, however, seems to be as far as bacteria have explored the opportunities provided by multicellularity.

Among eukaryotes, a similar level of complexity has independently been achieved many times. For example, volvocine algae have apparently evolved their version of multicellularity (involving the differentiation of reproductive and ciliated cells) several times independently. Note that the mechanism involved intrinsically limits the volvocine algae this very basic level of multicellularity. As the nearest relatives of metazoans, choanoflagellates are of particular interest; most choanoflagellates are unicellular, but some are not. In the case of *Salpingoeca*, while the mechanism of multicellularity is an elegant reflection of selective forces, the molecules and processes involved are irrelevant in terms of understanding animal origins. In response to the abundance of prey bacteria, the unicellular form of *Salpingoeca* forms aggregates that can better exploit the food source; the

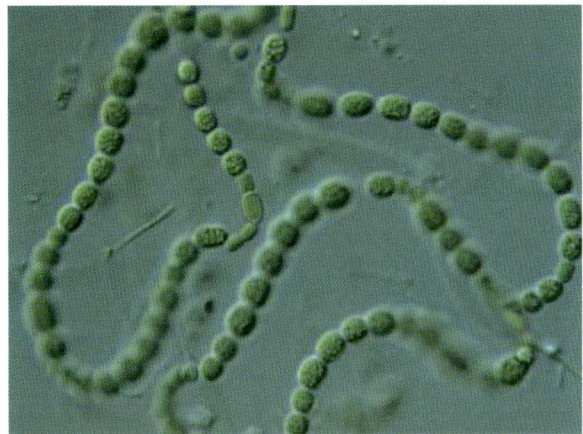

Fig. 2.6 The nitrogen-fixing bacterium *Nostoc* is a commonly occurring cyanobacterium often found in symbiotic associations (Taken from proprofs.com)

mechanism involves recognition by the choanoflagellate of a specific sulfolipid in the cell wall of a restricted group of bacteria.

This kind of multicellularity is obviously very different both qualitatively and quantitatively from that of animals, but it is important to recognize that intermediate levels of complexity exist. Green plants have achieved a comparable level of complexity to that of animals, with fungi and the red and brown algal lineages at a somewhat less advanced level. That despite multicellularity per se being a relatively simple trick to perform, there have been just five successful independent ventures into "advanced" multicellularity in over 3.5 billion years of opportunity that indicates that the events or innovations involved in founding these lineages were of a fundamentally different nature, that leading to the Metazoa being perhaps the most significant in terms of the evolution of the biosphere.

2.8 Multicellularity Requires Cooperation of Cells

Below, we will consider possible genetic drivers of the Cambrian and earlier animal diversifications, but first, with so much of a head start, why have prokaryotes failed to evolve beyond a very basic version of multicellularity? One possibility is that the nature of bacterial evolution itself may serve to effectively limit them to this basic level of multicellularity. Cellular specialization demands integration of cells, with greater integration being required for higher levels of multicellularity. However, integration requires cooperation of cells, which means that each cell must be able to "trust" its neighbors—they must operate in fundamentally similar ways, otherwise, for example, effective communication between cells is not possible, and neighbors could "cheat" rather than cooperate in the multicellular endeavor. Multicellularity therefore is contingent on genetic homogeneity—cells can only effectively communicate with ("trust") its neighbors which have essentially the same genotype. Genetic homogeneity is a hallmark of multicellularity, but is probably quite rare in prokaryotes.

On the contrary, bacteria are genetically promiscuous, exchanging genes with remarkable fluidity, while this facilitates rapid evolution to changing environmental conditions, it limits the ability of bacterial cells to cooperate and integrate and hence achieve complex multicellularity. One corollary of this line of thinking is that the evolution of more complex forms multicellularity should require the evolution of increasingly sophisticated ways of protecting the genome from invading foreign DNA. Both animals and plants do have formidable molecular arsenals that can be deployed against invading foreign DNA, which accounts for the rarity of lateral gene transfers (LGTs) into their genomes. In fact, many putative LGTs into animal genomes are more likely to reflect random but widespread gene loss from a genetically complex ancestor.

The case for very early origins for extant animals is based on molecular phylogenetic arguments and comparing fossil data with extant animals. Molecular phylogenetic approaches assume that it is possible to extrapolate recent rates of sequence evolution back to deep time, but such extrapolations often come with enormous standard deviations, and the nature of the analyses could violate the basic assumptions of molecular phylogenetic approaches. The interpretation of Precambrian fossils is at least equally contentious, with alarming frequency, new discovered early Precambrian fossils or new interpretations of known materials are claimed to either push back the origins of the Metazoa or confirm deep Precambrian origins, the papers (often in high-impact journals) only to be refuted shortly thereafter. Here, we take the conservative view that the Metazoa arose at or close to the Ediacaran/Cambrian boundary and that all or most of the Precambrian fossils represent something else entirely. Before the late Ediacaran, animal life was scarce, simple, and restricted in distribution, but then suddenly the entire game changed—animals were abundant, complex, and more widely distributed and began to have a much greater impact on the biosphere.

2.9 Genomes of Early Emerging Metazoans Reveal the Origin of Animal-Specific Genes

During animal development, formation of morphologically complex structures from the fertilized egg requires precise cell-to-cell communication. Surprisingly, only a handful of intercellular signaling pathways are used throughout the animal kingdom: wnt, tgf-ß, hedgehog, receptor tyrosine kinase, and notch. The genome of a marine sponge, *Amphimedon queenslandica* (Fig. 2.5), has provided fascinating insights into the origins of animal evolution (Srivastava et al. 2010). The *A. queenslandica* genome harbors an extensive repertoire of developmental signaling and transcription factor genes, indicating that the metazoan ancestor had a developmental "toolkit" similar to that in modern complex bilaterians. Analysis of the *Amphimedon* genome and expression studies done by Maja Adamska's team at the SARS Center in Bergen (Adamska et al. 2010; Adamska et al. 2011; Fortunato et al. 2014) demonstrate that all major metazoan signaling pathways are used during sponge embryogenesis. Wnt, tgf-ß, and hedgling—related to hedgehog transmembrane protein—genes are expressed in dynamic, partially overlapping patterns in *Amphimedon* embryos.

2.9 Genomes of Early Emerging Metazoans Reveal the Origin of Animal-Specific Genes

Similarly to higher animals, the wnt pathway appears to be used in establishment of the anterior–posterior axis of the sponge embryo. The origins of many of these and other genes specific to animal processes such as cell adhesion, and social control of cell proliferation, death, and differentiation can be traced to genomic events (gene birth, subfamily expansions, intron gain/loss, and so on) that occurred in the lineage that led to the metazoan ancestor, after animals diverged from their unicellular "cousins." In addition to possessing a wide range of metazoan-specific genes, the *Amphimedon* draft genome is missing some genes that are conserved in other animals, indicative of gene origin and expansion in eumetazoans after their divergence from the demosponge lineage and/or gene loss in *Amphimedon*.

While the sponge and cnidarian data pushed back the evolutionary origins of developmental signaling pathways and many transcription factors to common metazoan (or eumetazoan) ancestores (Shinzato et al. 2011), it is now clear that many "animal-specific" genes have older origins yet and are not restricted to animals. Clear evidence for more widespread occurrence of "animal-specific" genes came from sequencing of transcripts and then the genome of the choanoflagellate, *Monosiga brevicollis*. Choanoflagellates promise to provide insights into the origins of multicellularity because they sometimes form colonies in which cells maintain their individuality but can also share small molecules. Nicole King's lab at UC Berkeley has shown that choanoflagellate genomes do encode a number of "animal-specific" genes, including receptor tyrosine kinases, cadherins, and immunoglobulins (Alegado and King 2014; Alegado et al. 2012). Disappointingly, sequencing the genome of the "multicellular" choanoflagellate, *Salpingoeca rosetta* (Fig. 2.7), added very little to understanding metazoan origins; it seems that choanoflagellates have independently invented a mechanism (or mechanisms) for "multicellularity." Thus, although they tell us more about how multicellularity can be achieved, choanoflagellates cannot inform us about how the metazoan ancestor made this transition.

The recent focus of genome sequencing efforts on other unicellular holozoans highlights just how old some "animal specific" genes actually are. Surprisingly, the genome of the filasterean holozoan *Capsaspora*, which may have been secondarily reduced given its endosymbiotic (or parasitic) lifestyle, encodes a number of "metazoan" transcription factors that have been lost from choanoflagellates, including the RUNX, NFk, and the T-box protein Brachyury. As efforts to sequence the genomes of a truly representative range of close outgroups and early diverging ingroups of the Metazoa continue, it is likely that the list of animal-specific genes will continue to shrink. In reality, only a small number of gene families or protein domains are likely to be restricted to the animal kingdom.

Although stochastic and widespread, gene loss accounts for many putative LGTs into animal genomes, a few cases remain of genes typical of other kingdoms being anomalously distributed in only one (or, at most a few) metazoan lineage(s). Classic examples include the cellulose biosynthetic genes of ascidians, whereas cellulose is a common cell wall constituent in plants, fungi, slime molds, and bacteria, among animals, it is only known from the tunics of ascidians. This distribution gave rise to the idea that the biosynthetic pathway was acquired in ascidians by "horizontal transfer" (= LGT). However, if the genes were acquired by LGT,

Fig. 2.7 A cluster of colony-forming *Salpingoeca rosetta* choanoflagellates. Triggered by the presence of bacteria, the single-celled choanoflagellate *Salpingoeca rosetta* divides and stays with its sisters to form a colony. Scanning electron microscope image courtesy of Nicole King/news.berkeley.ed

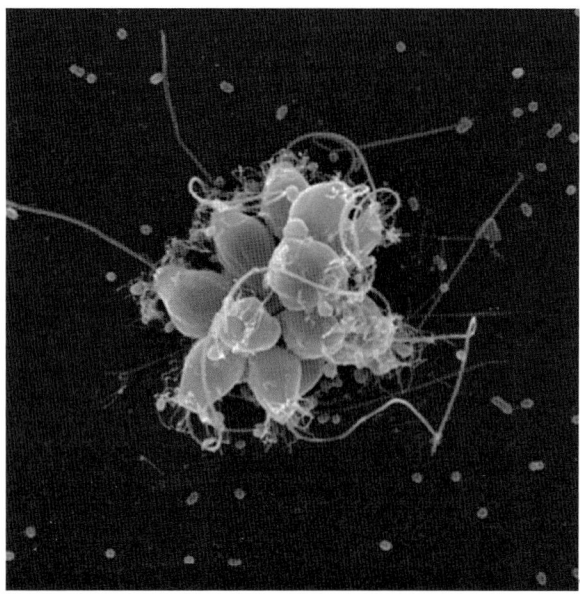

this happened a long time ago—much longer ago than is usually implied when LGT is invoked—because cellulose synthase genes from ascidians that diverged deep in the Cambrian are very similar. Rather than a recent LGT, this pushes the time of any "horizontal transfer" of cellulose synthase genes to very early in metazoan evolution. Similar scenarios apply to a number of other animal genes with anomalous distributions.

The *Hydra magnipapillata* genome was sequenced in 2010 by the Sanger whole genome shotgun method at the J. Craig Venter Institute (Chapman et al. 2010). Following the completion of sequencing, two assemblies of the genome were done. The assembly gives an estimated nonredundant genome size of 1.05 Gb. The *Hydra* genome is (A+T)-rich and contains ~20,000 bona fide protein-coding genes. All the essential signaling pathways controlling the formation of epithelia, muscle tissue, stem cells, nerve cells, and the innate immune system are present in the *Hydra* genome.

All these observations in choanoflagellates, sponges, and cnidarians together make not only a very strong point that the "molecular tool kit" leading to the evolution of higher animals and humans was already present at the beginning of multicellular life. They also show that animal life from the very beginning was multiorganismal and involved close association between simple animal hosts, bacteria, and in many cases also eukaryotic symbionts and viruses (for reference see, e.g., King 2010).

As we will see next, embedding this diversity of animal life within a clear phylogenetic framework is essential to understanding it from both molecular biology and evolutionary perspective.

References

Adamska M, Larroux C, Adamski M, Green K, Lovas E, Koop D, Richards GS, Zwafink C, Degnan BM (2010) Structure and expression of conserved Wnt pathway components in the demosponge Amphimedon queenslandica. Evol Dev 12(5):494–518

Adamska M, Degnan BM, Green K, Zwafink C (2011) What sponges can tell us about the evolution of developmental processes. Zoology (Jena) 114(1):1–10, Review

Alegado RA, King N (2014) Bacterial influences on animal origins. Cold Spring Harb Perspect Biol 6:a016162

Alegado RA, Brown LW, Cao S, Dermenjian RK, Zuzow R, Fairclough SR, Clardy J, King N (2012) A bacterial sulfonolipid triggers multicellular development in the closest living relatives of animals. ELife 1:e00013

Artamonova II, Mushegiand AR (2013) Genome sequence analysis indicates that the model eukaryote nematostella vectensis harbors bacterial consorts. Appl Environ Microbiol 79(22):6868–6873

Barrangou R, Fremaux C, Deveau H, Richards M, Boyaval P, Moineau S, Romero DA, Horvath P (2007) CRISPR provides acquired resistance against viruses in prokaryotes. Science 315:1709–1712

Chapman JA, Kirkness EF, Simakov O, Hampson SE, Mitros T, Weinmaier T, Rattei T, Balasubramanian PG, Borman J, Busam D, Disbennett K, Pfannkoch C, Sumin N, Sutton GG, Viswanathan LD, Walenz B, Goodstein DM, Hellsten U, Kawashima T, Prochnik SE, Putnam NH, Shu S, Blumberg B, Dana CE, Gee L, Kibler DF, Law L, Lindgens D, Martinez DE, Peng J, Wigge PA, Bertulat B, Guder C, Nakamura Y, Ozbek S, Watanabe H, Khalturin K, Hemmrich G, Franke A, Augustin R, Fraune S, Hayakawa E, Hayakawa S, Hirose M, Hwang JS, Ikeo K, Nishimiya-Fujisawa C, Ogura A, Takahashi T, Steinmetz PR, Zhang X, Aufschnaiter R, Eder MK, Gorny AK, Salvenmoser W, Heimberg AM, Wheeler BM, Peterson KJ, Bottger A, Tischler P, Wolf A, Gojobori T, Remington KA, Strausberg RL, Venter JC, Technau U, Hobmayer B, Bosch TC, Holstein TW, Fujisawa T, Bode HR, David CN, Rokhsar DS, Steele RE (2010) The dynamic genome of Hydra. Nature 464(7288):592–596

Doolittle WF, Fraser P, Gerstein MB, Graveley BR, Henikoff S, Huttenhower C, Oshlack A, Ponting CP, Rinn JL, Schatz MC, Ule J, Weigel D, Weinstock GM (2013) Sixty years of genome biology. Genome Biol 14(4):113

Doolittle WF, Zhaxybayeva O (2009) On the origin of prokaryotic species. Genome Res 19(5):744–756

Fortunato SAV, Adamski M, Mendivil Ramos O, Leininger S, Liu J, Ferrier DEK, Adamska M (2014) Calcisponges have a ParaHox gene and dynamic expression of dispersed NK homeobox genes. Nature 514:620–623

Fukami T, Wardle DA, Bellingham PJ, Mulder CP, Towns DR, Yeates GW, Bonner KI, Durrett MS, Grant-Hoffman MN, Williamson WM (2006) Above- and below-ground impacts of introduced predators in seabird-dominated island ecosystems. Ecol Lett 9:1299–1307

Grissa I, Vergnaud G, Pourcel C (2007) The CRISPRdb database and tools to display CRISPRs and to generate dictionaries of spacers and repeats. BMC Bioinf 23(8):172

Hadfield MG (2011) Biofilms and marine invertebrate larvae: what bacteria produce that larvae use to choose settlement sites. Ann Rev Mar Sci 3:453–470

Horvath P, Barrangou R (2010) CRISPR/Cas, the immune system of bacteria and archaea. Science 327:167–170

Keeling PJ, McCutcheon JP, Doolittle WF (2015) Symbiosis becoming permanent: survival of the luckiest. Proc Natl Acad Sci U S A 112(33):10101–10103

King N (2010) Nature and nurture in the evolution of cell biology. Mol Biol Cell 21:3801–3802

Knoll AH (2003) Life on a young planet. Princeton University Press, Princeton

Makarova KS, Haft DH, Barrangou R, Brouns SJJ, Charpentier E, Horvath P, Moineau S, Mojica FJM, Wolf YI, Yakunin AF, Oost J, Koonin EV (2011) Evolution and classification of the CRISPR-Cas systems. Nat Rev Microbiol 9:467–477

Margulis L, Dorion S (2001) Marvellous microbes. Resurgence 206:10–12

McFall-Ngai M, Hadfield M, Bosch T, Carey H, Domazet-Loso T, Douglas A, Dubilier N, Eberl G, Fukami T, Gilbert S, Hentschel U, King N, Kjelleberg S, Knoll A, Kremer N, Mazmanian S, Metcalf J, Nealson K, Pierce N, Rawls J, Reid A, Ruby E, Rumpho M, Sanders J, Tautz D, Wernegreen J (2013) Animals in a bacterial world, a new imperative for the life sciences. Proc Natl Acad Sci U S A 110(9):3229–3236

Shinzato C, Shoguchi E, Kawashima T, Hamada M, Hisata K, Tanaka M et al (2011) Using the Acropora digitifera genome to understand coral responses to environmental change. Nature 476:320–323

Shoguchi E, Shinzato C, Kawashima T, Gyoja F, Mungpakdee S, Koyanagi R, Takeuchi T, Hisata K, Tanaka M, Fujiwara M, Hamada M, Seidi A, Fujie M, Usami T, Goto H, Yamasaki S, Arakaki N, Suzuki Y, Sugano S, Toyoda A, Kuroki Y, Fujiyama A, Medina M, Coffroth MA, Bhattacharya D, Sato N (2013) Draft assembly of the symbiodinium minutum nuclear genome reveals dinoflagellate gene structure. Curr Biol 23:1399–1408

Srivastava M, Simakov O, Chapman J, Fahey B, Gauthier MEA, Mitros T, Richards GS, Conaco C, Dacre M, Hellsten U, Larroux C, Putnam NH, Stanke M, Adamska M, Darling A, Degnan SM, Oakley TH, Plachetzki DC, Zhai Y, Adamski M, Calcino A, Cummins SF, Goodstein DM, Harris C, Jackson DJ, Leys SP, Shu S, Woodcroft BJ, Vervoort M, Kosik KS, Manning G, Degnan BM, Rokhsar DS (2010) The Amphimedon queenslandica genome and the evolution of animal complexity. Nature 466(7307):720–726

Taylor LH, Latham SM, Woolhouse ME (2001) Risk factors for human disease emergence. Philos Trans R Soc Lond B Biol Sci 356(1411):983–989

Williams GP, Babu S, Ravikumar S, Kathiresan K, Prathap SA, Chinnapparaj S, Marian MP, Alikhan SL (2007) Antimicrobial activity of tissue and associated bacteria from benthic sea anemone *Stichodactyla haddoni* against microbial pathogens. J Environ Biol 28:789–793

Woese CR, Fox GE (1977) Phylogenetic structure of the prokaryotic domain: the primary kingdoms. Proc Natl Acad Sci U S A 74(11):5088–5090

The Diversity of Animal Life: Introduction to Early Emerging Metazoans

3

Extant animals are classified into about 26 phyla, each of which capture variations in a basic body plan (Bauplan) that, with only one exception, dates back to the Cambrian. On the basis of molecular criteria, most of these are grouped into three "superphyla" (Fig. 3.1)—Ecdysozoa (the "moulting" animals, within which the arthropods and nematodes are the major component phyla), Lophotrochozoa (crest or wheel animals; mollusks, annelids, and platyhelminths are the major groups), and Deuterostomia (echinoderms, hemichordates, urochordates, and chordates). The overwhelming majority of extant animals have two obvious body axes—anterior/posterior and dorsal/ventral—and hence the phyla to which they belong are known as the Bilateria (or bilaterally symmetric animals), the "higher" animals. That the Bilateria have a single origin (i.e., are a monophyletic group) is most convincingly shown by the fact that the molecular mechanisms involved in patterning along these two axes are conserved even between such different kinds of animals as vertebrates and insects.

There are, however, at least four extant phyla whose origins predate those of the Bilateria: these are the Cnidaria (jellyfish, sea anemones, corals, hydras), Porifera (sponges), Ctenophora (comb jellies), and Placozoa (no common name). Cnidaria and Porifera are large and diverse phyla, each containing many thousands of species, whereas the Ctenophora is a small group and the Placozoa a phylum of one. Whole-genome sequences are available for representatives of each of these, but their relationships are still equivocal and hotly debated (see also below). Perhaps even more contentious are the interrelated issues of the timing of the emergence of the Metazoa and the relationship between extant animals are those that are known only from Precambrian fossils.

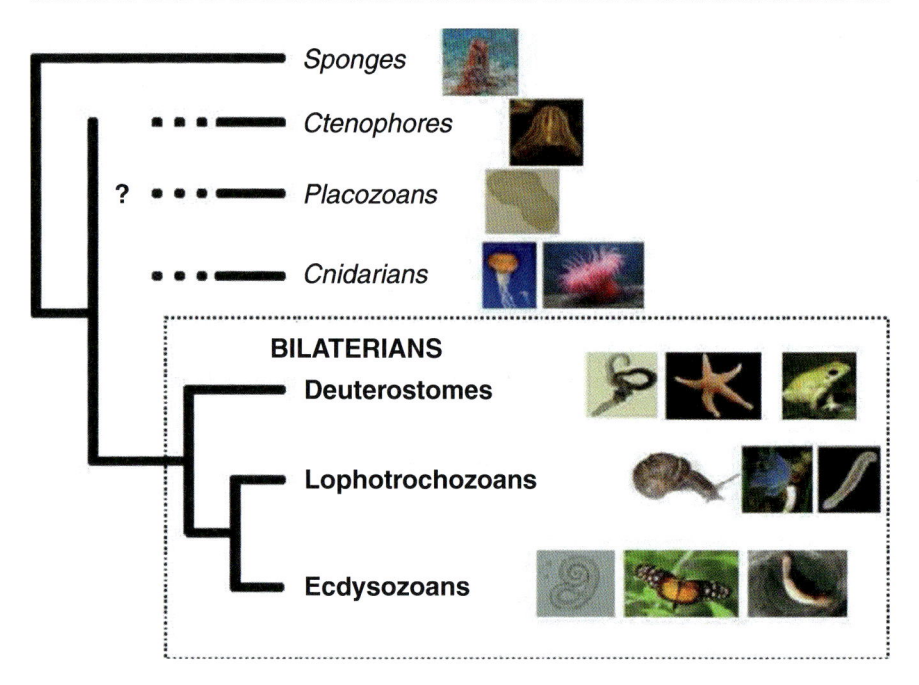

Fig. 3.1 Relationships between the major extant groups of animals. Three superphyla (*Deuterostomes, Lophotrochozoa*, and *Ecdysozoa*) comprise the Bilateria—the clade of "higher" animals. Despite the availability of whole-genome sequences for representatives of all of the "non-bilaterian" phyla, their relationships are equivocal, and the Porifera may actually comprise more than one phylum. The Bilateria are defined by the presence of two obvious axes of symmetry, whereas at least four phyla are nominally non-bilaterian; placozoans have an upper and lower surface, but no other axes of symmetry and adult sponges often have irregular patterns of organization, but their larvae (and the adults of some species) are radially organized (i.e., have a single axis of symmetry). Although cnidarians and ctenophores are traditionally regarded as having radial symmetry, ctenophores display a wide range of body types and true radial symmetry is very rare in cnidarians. Many cnidarians have a kind of bilateral symmetry as adults, and a cryptic secondary body axis is also revealed by gene expression patterns in larvae even in the absence of morphological correlates (see Fig. 3.5) (Figure source: https://www.cambrianmammal.wordpress.com/)

3.1 How Old Are the "Early Diverging" Animal Phyla?

Despite the long history of life on earth, familiar kinds of animals emerged suddenly and in spectacular variety at or close to the base of the Cambrian. The Cambrian "explosion" has been dated very precisely at around 543 MYA, and the great diversification happened within a very narrow time window of 20–25 MY. Animals of some kind were undoubtedly present during the preceding Ediacaran Period (635–542 MYA), but these were much less abundant and typically do not resemble any extant lineages. Some examples of fossils from the early Cambrian and the preceding Ediacaran are shown in Fig. 3.2. Many of the Ediacaran fossils are bizarre, wonderful, and completely enigmatic. That some of them are of biological origin at all has been questioned, but there is near unanimous agreement that most are. The known Ediacaran fossils represent about 6 distinct kinds of body plans. Some

3.1 How Old Are the "Early Diverging" Animal Phyla?

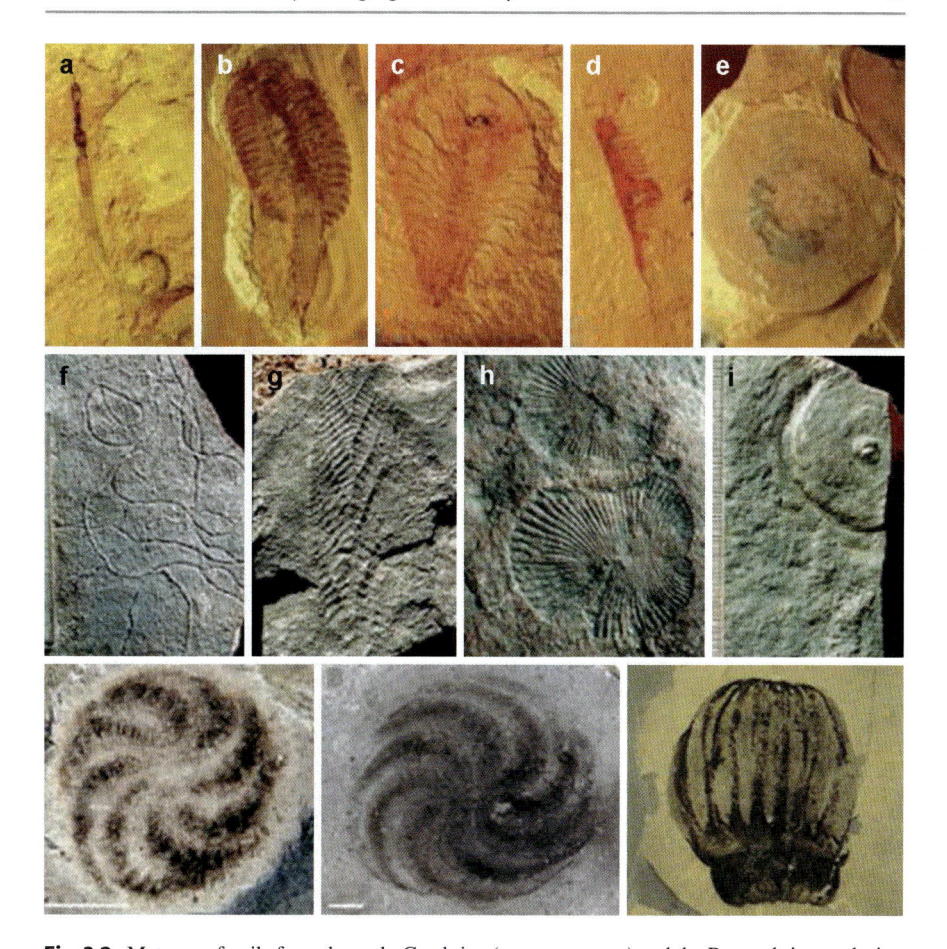

Fig. 3.2 Metazoan fossils from the early Cambrian (*upper row*, **a–e**) and the Precambrian explosion Ediacaran (*lower row*, **f–i**). Whereas many of the early Cambrian fossils (*top line*) can be interpreted in terms of extant phyla or their shared ancestors, for the Ediacaran fossils (*middle panel*) this is often much more problematic. (**a**) *Maotianshania*, thought to be a priapulid. (**b**) *Fuxianhuia* and (**c**) *Xandarella*, considered to be primitive arthropods. (**d**) *Jiangfengia*, a "great appendage" arthropod. (**e**) *Eldonia*, a pelagic Ediacaran—note the curved digestive tract. Forms **a–e**, from the Chengjiang Formation, Yunnan, China, are between 1.7 and 6 cm in greatest dimension. (**f**) Helminthoidichnites, fossil trails; Blueflower Formation, Mackenzie Mountains, northwestern Canada. (**g**) *Charnia masoni*, was; Ust Pinega Formation, White Sea, Russia. The relatively abundant *Dickinsonia* (**h**) has been variously interpreted as a lichen, fungus, cnidarian, annelid, or placozoan, but there are problems with all of these interpretations; Ediacara member, Rawnsley quartzite, Australia. (**i**) *Ediacaria*, a benthic cnidarian-like form, basal view to show stem, Sheepbed Formation, Mackenzie Mountains, northwestern Canada. Body fossils of these forms can be large; *Dickinsonia* may reach 1 m. (Source of images **a–i**: Valentine et al. (1999)). The lower row of images is of fossils that have been interpreted as ctenophores. The two images at left show *Eoandromida*, an Ediacaran fossil interpreted as a ctenophore by some authors but only distantly resembling extant representatives, whereas viewing *Ctenorhabdotus* from the Burgess Shale (*right image*) as a stem group ("primitive") ctenophore is far less. Note that, whereas extant ctenophores typically possess 8 comb rows, *Ctenorhabdotus* had 24 (Images are from: Tang et al. 2011 *Evol Dev* 13, 408–414 and http://www.burgess-shale.rom.on.ca/en/science/burgess-shale/03-fossils.php#box9)

of the Ediacaran forms persisted into the Cambrian, but by its end all were extinct. Some of the "Vendobiota" (the Vendian is a term that has largely been superseded by the term Ediacaran) were christened the "quilted organisms" by Seilacher, because they resemble inflatable mattresses. Quilted organisms such as *Charnia* were stalked and must have filter-fed on the bacterial soup, while others such as *Dickinsonia* were unattached. The latter certainly seems to have used cilia for feeding, presumably drifting over and browsing on the microbial mat. These organisms resemble nothing that exists today, but some Ediacaran fossils at least superficially are much more reminiscent of modern animals. With its regular lateral branches, *Charniodiscus* resembles a modern sea pen (Fig. 3.3). However, this interpretation is likely to be correct. Of the 32 genera of extant pennatulaceans (sea pens), only six (Williams 2011) have the large, well-developed polyp leaves that superficially resemble those of the Ediacaran forms. These six taxa—which include *Ptilosarcus*—are considered the most highly derived types of sea pens (Williams 1994) and are not known from any geological period prior to the Tertiary (62.5–2.6 MYA). The Ediacaran forms are for the most part foliate (leaf-like) and have a continuous margin around the leaf-like frond. The extant sea pens with polyp leaves, on the other hand, are pennate (feather-like), having numerous lateral appendages attached only at their bases to the central rachis. The lateral appendages of all of these extant taxa are actually separate polyp leaves in two opposite rows along the rachis, containing the polyps. The single frond of the Ediacaran forms cannot be considered homologous or even functionally convergent with the polyp leaves of extant sea pens. Based on the paleontological and phylogenetic evidence, as well as that of comparative morphology, the Ediacaran/Burgess Shale "sea pens" are not fossilized pennatulaceans, but rather represent another unrelated but superficially similar lineage that was probably extinct by the end of the Cambrian. At present, there is no evidence to suggest that true pennatulaceans were present prior to the Mesozoic (252–66 MYA).

While a commonly held view is that the bilaterian phyla *except* Bryozoa (whose origins appear to have been early Ordovician or late Cambrian) have their origins in the early Cambrian, much earlier origins are often assumed or extrapolated for the non-bilateral phyla. In Fig. 3.4, hypothetical links between the extant bilaterian phyla are assumed by molecular clock-based extrapolation of sequence data and sometimes interpreting early fossils as representing precursors of one or more phyla. However, much deeper origins are frequently put forward for the Cnidaria and Porifera and sometimes likewise for the Ctenophora.

Some chemical data suggest that sponges might have very early origins, perhaps predating even the Ediacaran. The evidence for this was the detection of 24-isopropylcholestane, a molecule only previously known from demosponges, in Cryogenian deposits dating from at least 635 MYA. For some time the value of this putative biomarker has been questioned on the grounds of the absence of any hint of the fossil spicules that would be expected if demosponges really were present during the Precambrian—there is essentially a 200 million year gap between the nominal biomarker and the first fossil spicules, which are from the early Cambrian. More damaging to the idea of Precambrian sponge origins was the discovery of a 24-isopropyl steroid biosynthetic pathway in *Poribacter* (Siegl et al. 2011), a bacterium associated with marine sponges, implying that the nominal sponge biomarker

3.1 How Old Are the "Early Diverging" Animal Phyla?

Fig. 3.3 Similarities between Ediacaran fossils and extant forms can be misleading. *Charniodiscus* (*left*) is a relatively common frond-like fossil from the Ediacaran and early Cambrian that superficially resembles some modern sea pens, such as *Ptilosarcus* (*right*). However, morphological and phylogenetic evidence completely contradict this idea. *Charniodiscus* and other frond-like fossils from the Ediacaran and Burgess shale faunas (e.g.,, *Vaizitsinia*, *Khatyspytia*, *Thaumaptilon*) and extant leafy sea pens are considered to be neither homologous nor functionally convergent (*Charniodiscus* image: Xiao and Laflamme 2009. *Ptilosarcus* image: http://www.wrobelphoto.com/gallery/main.php?g2_itemId=704)

is actually a bacterial product and thus not informative with respect to metazoan origins.

Here we take the conservative view that the Metazoa—including the non-bilaterian phyla with which we are most concerned—arose at or close to the Ediacaran/Cambrian boundary and that the Precambrian fossils represent something else entirely. Some examples of fossils from the early Cambrian and the preceding Ediacaran are shown in Fig. 3.4. Whereas Cambrian fossils can generally (but not always) be interpreted without much difficulty in terms of extant body plans, many of the Ediacaran fossils defy attempts at classification. Before the late Ediacaran, animal life was scarce, simple, and restricted in distribution, but then suddenly the entire game changed—animals were more abundant, more complex, and more widely distributed and consequently began to have a much greater impact on the biosphere.

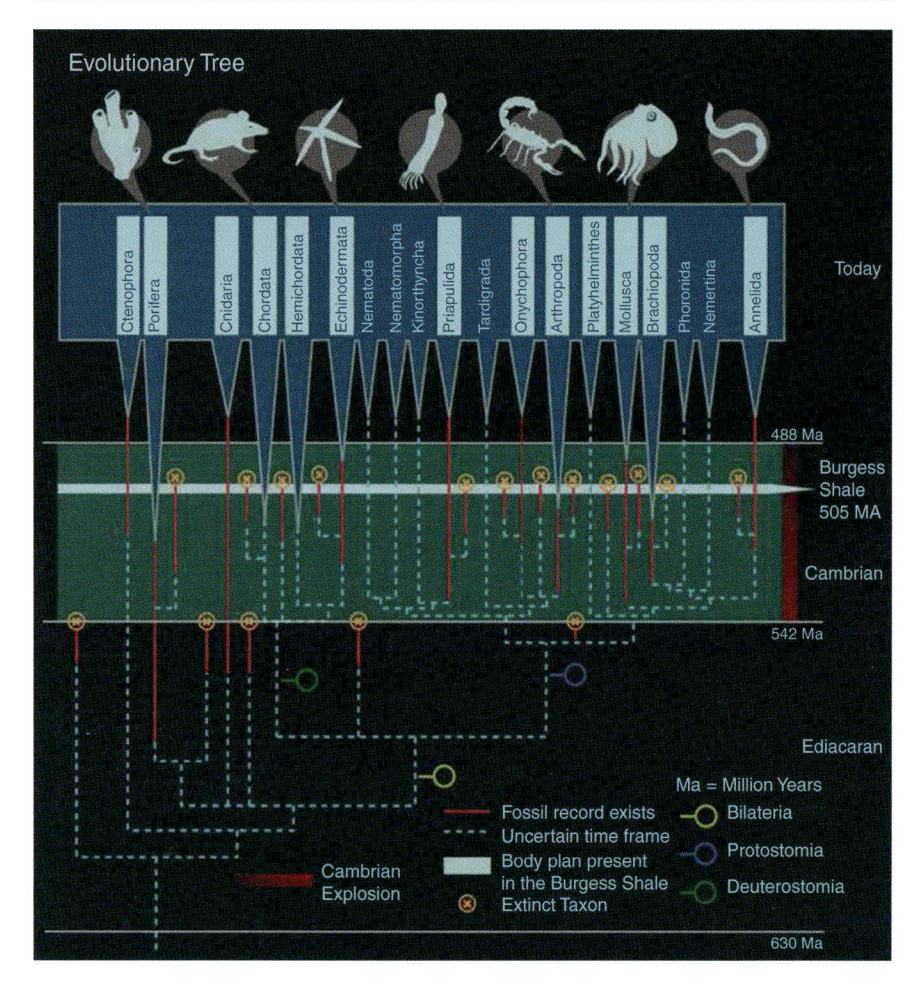

Fig. 3.4 Evolutionary origins of the major animal groups. Most of the extant phyla were distinct by the time of the Burgess Shale; dotted lines indicate their probable ranges. Solid lines indicate fossil evidence, and extinct groups are represented by a circled cross. The Precambrian part of this scheme is completely hypothetical; there is no compelling evidence for Ediacaran or earlier origins of the extant phyla Cnidaria and Porifera. Many of the Precambrian animals may have been at a similar grade of organization as cnidarians or sponges, but any similarities are superficial (see Fig. 3.3 for an example of this) (Image: http://www.burgess-shale.rom.on.ca/en/science/origin/04-cambrian-explosion.php)

3.2 Cnidarians: The Closest Relatives of "Higher" Animals (Bilateria)

Cnidaria is a large phylum, with around 9,000–10,000 species, comprising jellyfish, sea anemones, corals, and a host of other soft-bodied animals. Although some are found in freshwater—notably the textbook cnidarian, Hydra, but there are also a few freshwater jellyfish—the vast majority of cnidarians are marine. The phylum is defined by the possession of specialized structures known as cnidae, the major type of which is the nematocyst (or stinging apparatus). Nematocysts are barbed projectile tubes, through which toxins can be introduced and ejected at very high speed under mechanical stimulation. Nematocysts are used to immobilize and kill prey; in some cnidarians, other kinds of cnidae have specialized roles in entangling prey or, in the case of tube anemones (cerianthids), building the tube in which they live. The evolutionary success of the Cnidaria implies that the invention of cnidae was a major breakthrough, providing a major selective advantage to this lineage. Two broad types of life cycles are recognized across the four cnidarian classes; some have suggested that these might constitute two distinct phyla. In the "Medusozoa" (classes Scyphozoa, Cubozoa, and Hydrozoa), there is an alteration during the life cycle between a sedentary polyp phase and the sexually reproducing medusa (jellyfish) stage, whereas members of the large class Anthozoa lack the medusa stage, reproducing as polyps (Fig. 3.5).

While cnidarians are clearly animals, two major criteria have traditionally been used to distinguish them from "higher" animals (the Bilateria): the numbers of embryonic cell layers (three in bilaterians; two in cnidarians) and axes of symmetry. Cnidarians are often referred to as having radial symmetry along the single oral–aboral axis (i.e., having a single plane of symmetry), whereas by definition bilaterians have two axes of symmetry—anterior–posterior and dorsal–ventral. However, true radial symmetry is rare in cnidarians, most species showing at least minor deviations from this. In sea anemones, for example, the mouth is a slit, often with a ciliated groove at one end (the siphonoglyph), and the internal organization including the musculature is far from radially symmetric (Fig. 3.6). Gene expression data also support the presence of a cryptic secondary axis during early development of several cnidarians (Fig. 3.6. lower panel). Although ectoderm and endoderm are clearly present, cnidarians lack the third germ layer—mesoderm—in bilaterians; this gives rise to musculature, among other things. Some cnidarians (and one ctenophore) have clearly differentiated muscle types, but the origins of muscle differ radically from those known in bilaterians (see below).

There is a long history of cnidarian research stretching back to Abraham Trembley's work on *Hydra* regeneration more than 250 years ago. Consequently, of the four non-bilaterian phyla listed above, this is certainly the one that most is known about. The first whole-genome sequence for a non-bilateral animal was that of a cnidarian—the sea anemone *Nematostella vectensis* (Putnam et al. 2007), with those of *Hydra magnipapillata* (Chapman et al. 2010) and the coral *Acropora digitifera* (Shinzato et al. 2011) to follow a little later. Analyses of these confirmed the counterintuitive conclusion of earlier EST-based studies (Kortschak et al. 2003; Technau et al. 2005) that the common ancestor of Cnidaria and the Bilateria (the Ureumetazoan) possessed a surprisingly complex and vertebrate-like gene set that included most of the diversity of signaling pathways and transcription factors present in higher animals. Comparisons based on genome data for cnidarians and representatives of the other three pre-bilaterian phyla unequivocally support the idea that the Cnidaria are the closest relatives of the Bilateria; in the case of Cnidaria, evolutionary debates are mainly around what features are ancestral and internal, rather

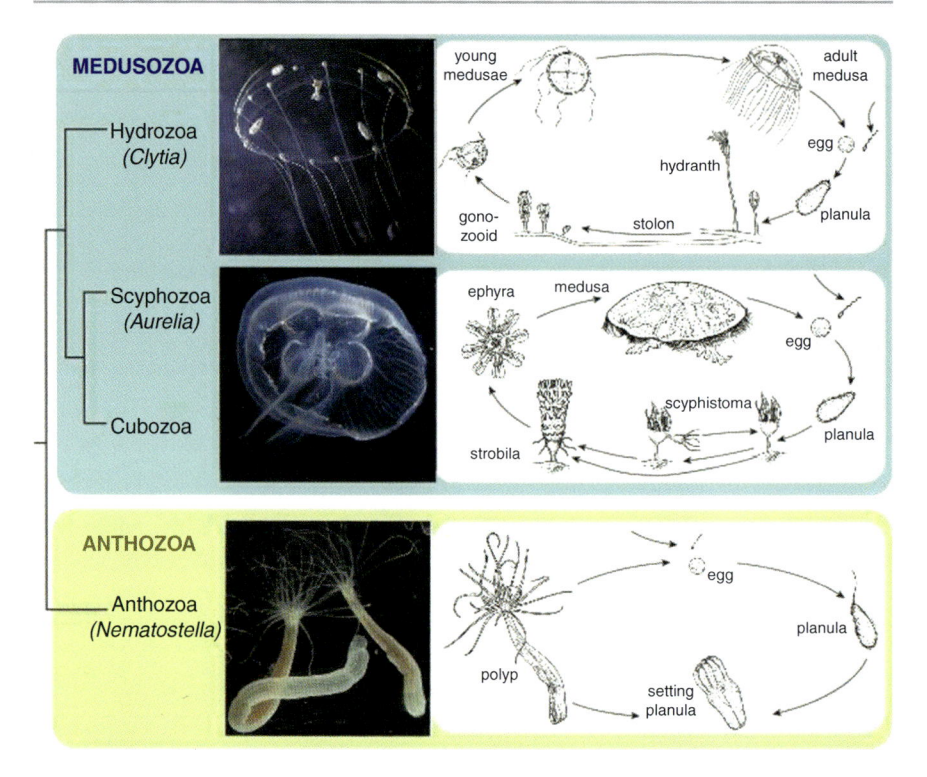

Fig. 3.5 Life cycles of representative cnidarians. Within the phylum, Anthozoa, which includes *Nematostella* (shown here) and the coral *Acropora*, is generally viewed as the earliest diverging. Whereas anthozoans have a "direct" life cycle, whereby the adult polyp reproduces via planula larvae, members of the other three classes, collectively known as the medusozoa, typically have a medusa or jellyfish stage as well as the polyp form. In medusozoans, the jellyfish is the sexually reproducing stage and is asexually derived from the polyp form. The figure shows *Aurelia aurita* and *Clytia hemispherica* as representative scyphozoans and hydrozoans, respectively. Hydra is also a hydrozoan but has secondarily lost the medusa stage and undergone extensive gene loss and sequence divergence. One outstanding question with respect to medusozoans is how the very different medusa and polyp stages are specified by a single genome (Figure is from Miller and Ball 2008)

than external, relationships. For example, one outstanding question is how the alternation between polyp and medusa ("jellyfish") life cycle stages (the asexual and sexually reproducing forms) is achieved in the "Medusozoa" (Fig. 3.5). Note that *Hydra* is an atypical hydrozoan that has secondarily lost the medusa stage; consequently the available genome data do not allow this question to be addressed.

3.3 Sponges: One Phylum or More?

Whereas there is broad agreement on the evolutionary position of cnidarians, in the case of sponges, the only unequivocal issue is that they diverged before the Cnidaria (Fig. 3.7). Whereas the long-standing assumption has been that the Porifera

3.3 Sponges: One Phylum or More?

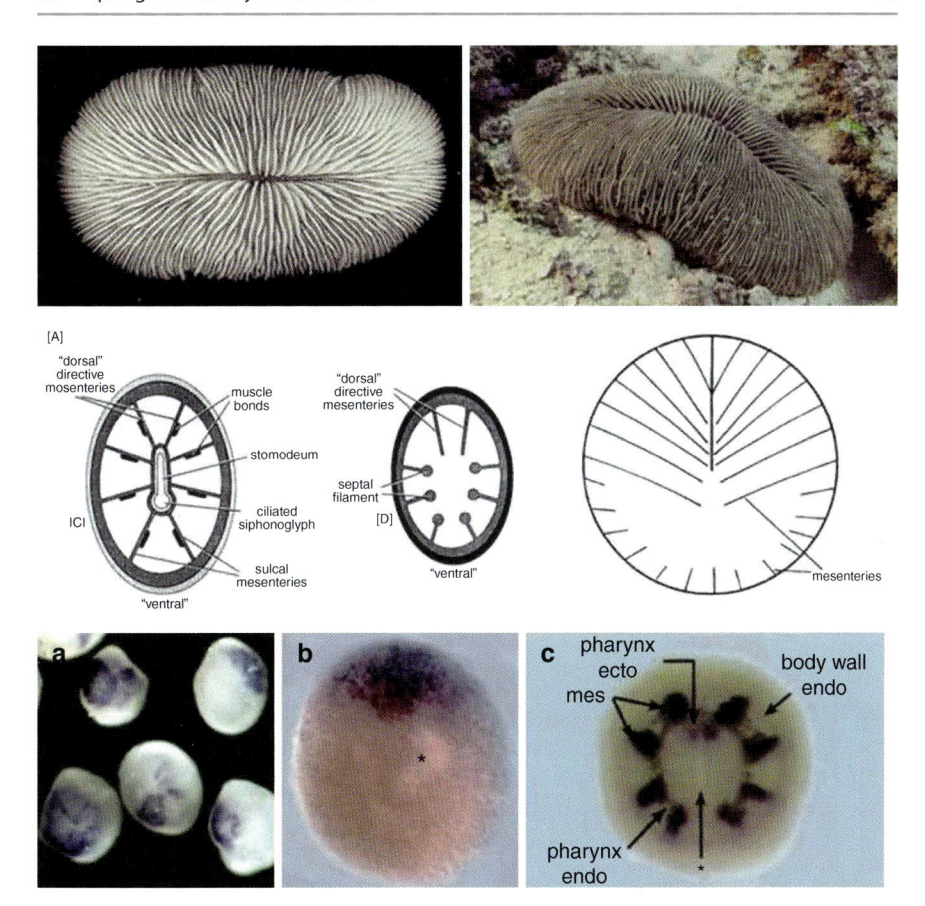

Fig. 3.6 Bilaterality in "radial" animals. (*Upper panels*) *Ctenactis*, one of the mushroom corals (family Fungiidae), typically has a large groove in the skeleton (*left*), which corresponds to the mouth of the adult (*right*). (*Middle panels*) The asymmetry across the primary (oral–aboral) axis can be clearly seen in the positioning of the siphonoglyph and mesenteries in sections of extant sea anemones (*left two panels*) and fossil corals (*right panel*). Images from Ball et al. 2007 Integrative and Comparative Biology 47: 701–711. (*Lower panels*) A cryptic second axis revealed by patterns of gene expression in anthozoan larvae. The cnidarian homolog of dpp/BMP2/4 is expressed in an asymmetrical patch at the edge of the blastopore (*) late in gastrulation in *Acropora millepora* (see figure part *b*) and *Nematostella vectensis* (*c*). There is only a single plane that can pass symmetrically through the area of gene expression and the blastopore, therefore fulfilling the definition of bilateral symmetry. Bilaterality is even more apparent in the polyp of *N. vectensis*, which has a slit-like mouth (*between the vertical arrows*), with dpp expression in the pharyngeal ectoderm at only one end (Images from: Ball et al. 2004)

represent the earliest diverging of extant metazoan phyla, there is a strong phylogenetic case for Ctenophora having earlier origins, and the true situation is now much less clear. One implication of a later divergence of the Porifera is that the structural simplicity that is a characteristic of this group may be secondary. The loss of complexity implied by this scenario is emerging as a recurring theme in metazoan

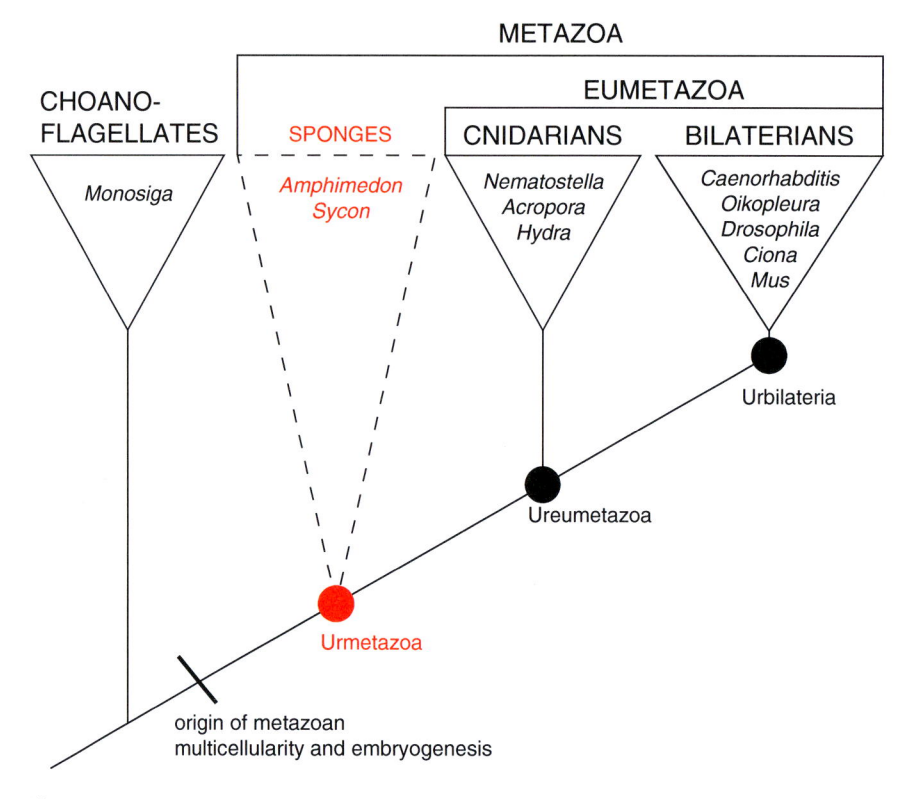

```
                              METAZOA
                                    EUMETAZOA
   CHOANO-
   FLAGELLATES    SPONGES    CNIDARIANS   BILATERIANS

     Monosiga    Amphimedon   Nematostella  Caenorhabditis
                  Sycon       Acropora      Oikopleura
                              Hydra         Drosophila
                                            Ciona
                                            Mus
```

Fig. 3.7 Sponges are considered to be the sister group of "true" animals, the Eumetazoa, and thus provide the ultimate outgroup for all comparative studies of animal development. Sponge monophyly remains a disputable issue (indicated by dashed lines) (With permission from Maja Adamska)

evolution—it is now clear that, at any level, complexity does not necessarily correlate with evolutionary divergence time.

Sponges are clearly at a less complex level of structural organization than are cnidarians; whereas in the latter case, cells are organized as germ layers, this is not the case in sponges. Sponges are "a collection of cell types covered by pinacoderm (a single cell layer) and supported by an exoskeleton consisting of spicules and/or spongin fibers." There are generally thin inner and outer "layers" of cells, but these differ from the true cell layers of eumetazoans in lacking a basement membrane (but see below). With very few exceptions, adult sponges are sessile filter feeders, trapping particulates by circulating water through often highly complex canal systems via the action of choanocytes—a specific kind of flagellated cells that is strikingly similar in overall appearance to the choanoflagellates, unicellular relatives of animals (Figs. 3.8 and 3.9).

Although adult sponges are sessile, their larvae are motile and planktonic and resemble the larvae of corals and other cnidarians. Although neither neurons nor a nervous system is present, sponges are capable of coordinated movement, but how this is achieved is unknown. Sponges are an incredibly diverse group of organisms

3.3 Sponges: One Phylum or More?

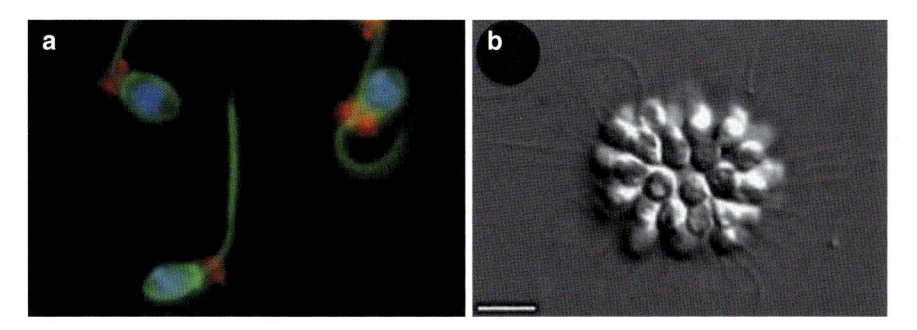

Fig. 3.8 Choanoflagellates resemble the choanocytes of sponges. Choanoflagellates such as *Monosiga brevicollis* (*upper left*) are unicellular holozoans whose structure is remarkably similar to the collar cells (choanocytes) of sponges. In both cases the flagellum is used to create water currents over the collar, enabling bacterial prey to be captured. Some choanoflagellates, such as *Salpingoeca rosetta* (*upper right*), can form colonies, although the mechanism involved in this kind of multicellularity is unique (Images: *Monosiga* (King 2005); *Salpingoeca* (Dayel 2011 Dev Biol 357, 73–82))

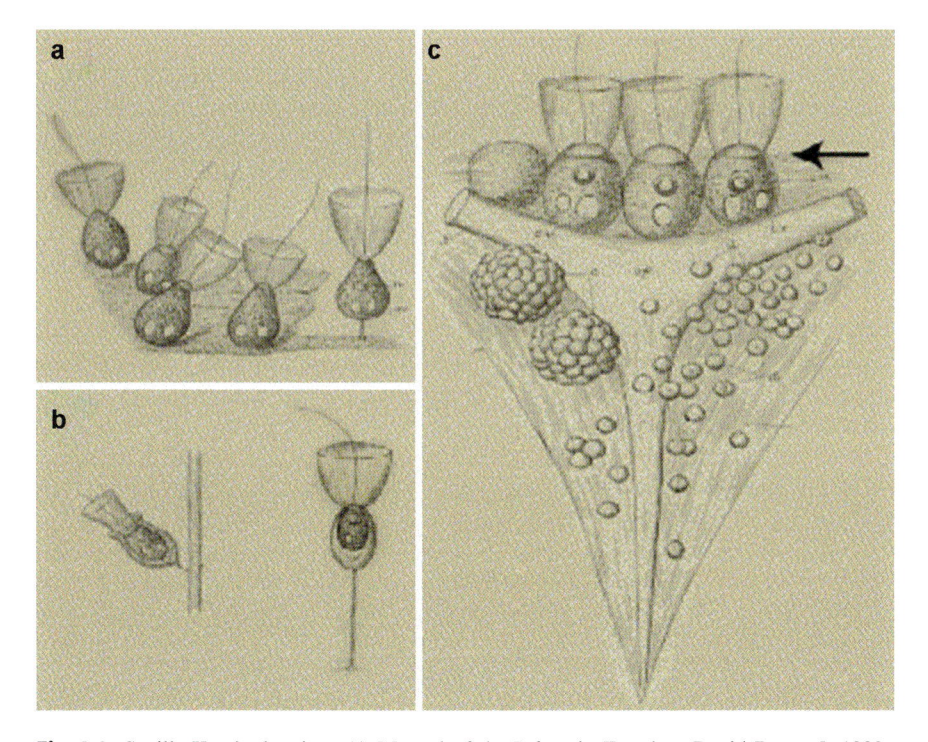

Fig. 3.9 Saville-Kent's drawings (A Manual of the Infusoria [London: David Brogue], 1880–1882) emphasize the similarities between choanoflagellates (**a** and **b**) and the choanocytes (**c**) of sponges; (**a**) *Monosiga consociata* (as modified from plate IV-19; Saville-Kent); (**b**) (*left*) *Salpingoeca convallaria* (as modified from plate IV-13; Saville-Kent), (*right*) *Salpingoeca infusionum* (as modified from plate VI-8; Saville-Kent); (**c**) *Leucosolenia coriacea*. Triradiate spicule (sp) and three associated choanocytes (*arrow*) (as modified from plate X-2; Saville-Kent) (Taken from (King 2004))

that fall into approximately four major groups that probably correspond to taxonomic classes—the demosponges (Demospongiae), homoscleromorphs, the glass sponges (Hexactinellida), and the calcareous sponges (Calcarea).

Of the 9–10 thousand extant species of sponges, the demosponges are by the far the largest group (85 %). This diverse Class is defined by the presence of the fibrous scleroprotein spongin—familiar as the material that forms the skeleton of bath sponges. Some demosponges also have mineralized elements—siliceous spicules or a calcareous basal skeleton—and in others the spongin skeleton has been largely replaced by collagen fibrils. Demosponges are ubiquitous in aquatic environments—in both fresh- and seawater, and from the depths to the shallows. The demosponges provide the "very few exceptions" to the poriferan norm mentioned above; members of the demosponge family Cladorhizidae (around 35–40 species) defy sponge norms in being carnivorous, this adaptation having led to a loss of typical sponge characteristics such as choanocyte chambers and a water circulatory system.

Homoscleromorpha is a small (fewer than 100 known species) but well-defined group of sponges, the external relationships of which are problematic. Most, but not all, members of the Homoscleromorpha occur in shallow waters and are typically smooth, encrusting sponges. Traditionally, homoscleromorphs were grouped with the demosponges, but they are now regarded as a distinct class. They are unique among sponges in the possession of basal lamina—in adult homoscleromorphs, the cell layers have basement membranes, which provide support and a site for attachment of other cells. The possession of basement membranes allows layers of cells to be segregated, enabling the organization of cells into tissues. Thus possession of basal lamina is a characteristic shared by homoscleromorphs and eumetazoans and which has traditionally been used to distinguish sponges from other animals. The homoscleromorph basal lamina resembles those of their eumetazoan counterparts in composition—for example, type IV collagens are present in both cases. On these bases, homoscleromorphs may either be basal within the sponge lineage or else constitute a distinct phylum.

The 600–650 known species of glass sponges (Hexactinellida) are restricted to marine habitats and, with very few exceptions, are found only in deep water. Glass sponges have siliceous skeletons in which the spicules have triaxonic symmetry, a characteristic not seen in any other sponges, and these are often highly elaborate structures. Another intriguing characteristic of glass sponges is that, as adults, they are largely syncytial—the cytoplasm of many cells is shared, rather than each cell being bounded by membrane layers.

Calcareous sponges (Calcarea; 650–700 species) are exclusively marine and occur mostly in shallow water. Although quite diverse in overall morphology, they are united in the positioning (but not the composition) of the spicules.

Calcareous spicules are also present in some representatives of other sponge classes but are always intracellular, whereas in calcareous sponges these are invariably extracellular. Although some calcareous sponges have skeletons consisting of fused spicules, these are in a minority—in most cases, the spicules are single entities.

3.4 The "Comb Jellies": The Enigmatic Phylum Ctenophora

The Ctenophora is a small phylum (150–200 known species), members of which are often referred to as the comb jellies on the basis of their obvious and defining characteristic; the "combs" are groups of fused cilia arranged in (usually 8) rows.

Ctenophores are at a similar level of structural complexity to cnidarians, which is to say that they are at a higher level of complexity than sponges. Ctenophores resemble cnidarians in having gelatinous and transparent bodies and in the majority of cases giving the appearance of near-radial symmetry. In both cases there is an obvious but non-centralized nervous system. These superficial similarities led to ctenophores and cnidarians historically being classified together as the Coelenterata. However, ctenophores lack the nematocysts or other cnidocyte derivatives that are the defining characteristic of cnidarians (although some ctenophores can brandish nematocytes derived from cnidarian prey) and many molecular characteristics clearly resolve the two phyla.

The ctenophore body consists of two epithelia (ectoderm and endoderm) covering a gelatinous mesoglea that is usually quite thick and often largely cellularized. The largely cellular structure of the ctenophore mesoglea contrasts with the situation in cnidarians, where in most cases the corresponding layer contains few cells, and historically sometimes led to it being misinterpreted as mesoderm. To further complicate matters, the mesoglea may also consist of relatively complex muscle—which, in bilaterians, is a mesodermal derivative. Hence, until recently, ctenophores were considered by some authors to be "higher" animals. Whereas motility in most animals is based on musculature, in ctenophores muscle is used primarily to support and provide shape for the gelatinous body rather than for locomotion. Some of the outer cells of the ectoderm are ciliated, enabling locomotion—ctenophores are the largest non-colonial animals that use cilia for locomotion (siphonophores, a group of hydrozoan cnidarians, are the largest animals to use cilia, but these are colonial). Colloblasts are a cell type unique to ctenophores; these are mushroom-shaped cells distributed on the ectoderm of the tentacles, which can produce a sticky substance which aids in capturing prey. Like the cnidarian nervous system, that of ctenophores is a nerve net with concentrations of neurons around the mouth, pharynx, and, in the latter, along the comb rows. A complex balance sensing statocyst is present in the aboral organ, which is the largest sensory organ of ctenophores. Cubozoan cnidarians have a similar balance sensor within the rhopalia, which in this case also provide "vision." Despite these apparent similarities, independent evolutionary origins have been suggested for the ctenophore and cnidarian/bilaterian nervous systems (see below).

Although the most familiar ctenophores—the sea gooseberries and sea walnuts (Fig. 3.10)—attain only modest sizes, some species can attain 1.5 m. The most familiar ctenophores are egg shaped and near radial in appearance, but dramatically different morphologies are known (Fig. 3.10). Members of the order Cestida are characteristically flattened in one plane, resembling ciliated ribbons (Fig. 3.10e). Platyctenids are perhaps the strangest of all ctenophores—these are also highly flattened, effectively creeping around on an everted stomodeum, and can be mistaken

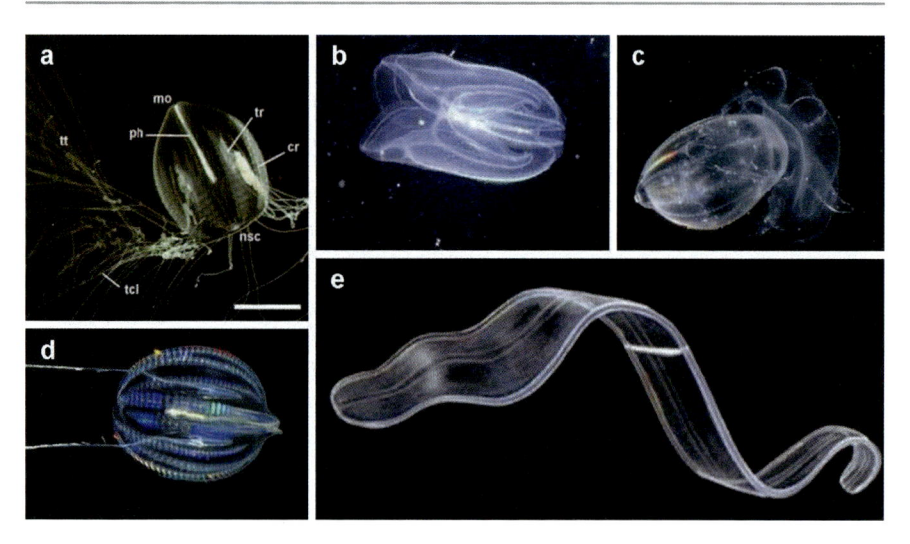

Fig. 3.10 Morphological diversity in the phylum Ctenophora. (**a**) The mouth (*mo*) and the aboral neurosensory complex (*nsc*) are located at the oral poles of the single body axis. The eight comb rows (*cr*) are locomotory structures, each made of regularly repeated swimming paddles or combs. The body is biradially symmetrical notably due to the presence of a single pair of tentacles (*tcl*), each bearing numerous lateral branchings or tentillae (*tt*). The tentacle root (*tr*), enclosed in the tentacle sheath, is responsible for constant regeneration of tentacle tissues. The pharynx (*ph*) and some canals of the endodermal gastrovascular system are also visible by transparency. Scale bar: 1 cm. From: Jager et al. 2013 PLoS One 8 e84363 (**b**) *Beroe ovata*, with its mouth open; (**c**) *Beroe* devours *Mnemiopsis*; (**d**) the sea gooseberry *Pleurobrachia bracei* (body size about 1.5 cm); (**e**) *Cestum veneris* (can reach 1.5 m long and around 5–8 cm wide) (combjellies.wikispaces.com)

for nudibrancs. Ctenophores are predators with prodigious appetites—when food is abundant, some can consume ten times their body mass in a day—and in a few cases the juvenile stages are parasites of the organisms on which the adult stages feed. *Beroe* is a particularly impressive predator, feeding on other ctenophores (such as *Mnemiopsis*) thanks to a large mouth with "teeth" that are actually hardened cilia. In Fig. 3.10b the gapping mouth of *Beroe ovata* can be seen, and Fig. 3.10c shows *Beroe* attacking *Mnemiopsis*.

3.5 Placozoans: The Simplest Extant Animals?

The adjective "enigmatic" is almost invariably attached to the anything relating to placozoans, which are small (a few millimeters across), flat, and irregularly shaped creatures consisting of two epithelia of cells sandwiching a fibrous interior. They were discovered in marine aquaria, feeding by external digestion carried out by the lower cell layer, moving by ciliary gliding and reproducing asexually by binary fission. Beyond having upper and lower surfaces relative to the substrate, placozoans have no trace of a body axis. In terms of cell type diversity, the placozoans are

3.5 Placozoans: The Simplest Extant Animals?

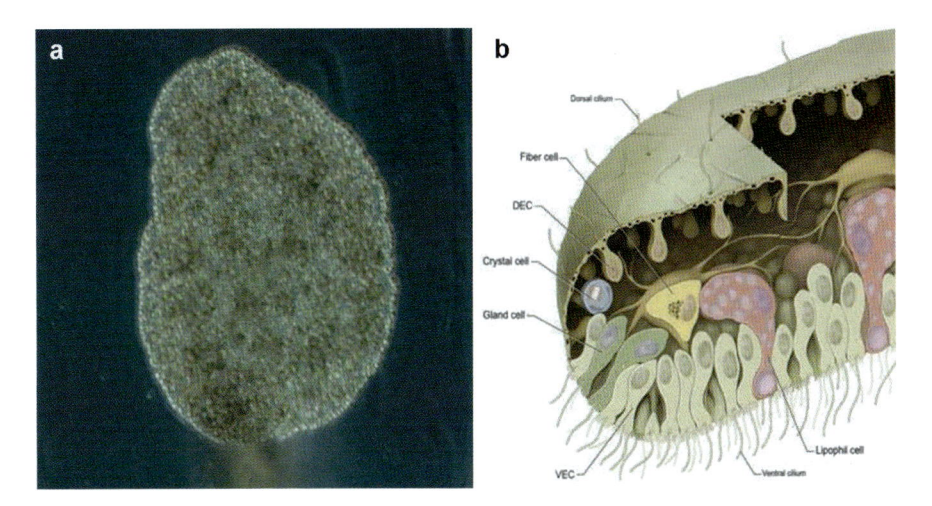

Fig. 3.11 The placozoan *Trichoplax adhaerens*. (**a**) Living *Trichoplax*. Placozoans are typically 0.5–2 mm (Photo: B Schierwater). (**b**) Schematic summary of the body plan of *Trichoplax* and its constituent cell types (From: Smith et al. 2014). A thickened lower surface layer composed of ventral epithelial cells (*VEC; light yellow*) faces the substrate. Close to the ventral surface, lipophil cells (*brick*) are present, containing lipid bodies, including a large spherical inclusion near the ventral surface (*lavender*). Gland cells (*pale green*) occur near the margin and are distinguished by the presence of secretory granules. The upper surface is composed of dorsal epithelial cells (*DEC; tan*), which form a roof across the top from which their cell bodies are suspended, surrounded by a fluid-filled space. Under the dorsal epithelium, a crystal cell (*pale blue*) containing a birefringent crystal is present near the margin. Fiber cells with branching processes are distributed between the dorsal and ventral epithelia and make contact with the other cell types

probably the simplest extant animals; at the time of release of the whole-genome assembly (Srivastava et al. 2008), only four cell types had been described, although a recent elegant structural study (Smith et al. 2014) has brought the cell type totally to six (Fig. 3.11).

The structural simplicity of placozoans led to the expectation that they might be the earliest diverging of all extant animals. While this idea still has some strong advocates, the consensus at the time of writing is that placozoans are secondarily simple and that they were probably derived from a cnidarian-like ancestor. The placozoan genome is the smallest among the known Metazoa, and the gene set is correspondingly small (around 10,500 genes) but the genes that remain are surprisingly bilaterian-like. The presence of what appear to be neurosecretory cells, a subset of which expresses a known neurotransmitter, is consistent with a more complex ancestry and otherwise difficult to rationalize in a simple animal that lacks a nervous system.

Placozoans are widely distributed throughout the oceans but have very limited morphological variation (Eitel et al. 2013). Only the single species *Trichoplax adhaerens* has been described, but there is a high level of underlying genetic diversity.

3.6 Eyes, Nervous Systems, and Muscles

The ability to rapidly respond to environmental cues is a typical animal characteristic and is also a characteristic of members of the non-bilaterian phyla Cnidaria and Ctenophora. Comparisons across the animal kingdom imply that the neurogenic machinery was largely in place in the common bilaterian ancestor and that some of its foundations were already present in the cnidarian/bilaterian ancestor.

Textbooks usually describe the nervous systems of cnidarians as diffuse nerve nets, and these are assumed to reflect the prototypes of bilaterian nervous systems and to have had common origins. Although it is certainly true that cnidarian nervous systems are often largely diffuse, in many cases dense masses of nerve cells or ganglia are also present. Despite its structural simplicity, the nervous system in cnidarians offers a rich chemical complexity. The ectodermal and endodermal nets consist of two types of neurons: peripheral sensory cells and sensory–motor interneurons. Muscle contraction is achieved via chemical synapses using large dense-cored vesicles containing a large variety of neuropeptides.

Moreover, although anthozoans (i.e., members of the earliest diverging class) lack obvious photoreceptors, some cubozoans have remarkably complex eyes (Kozmik et al. 2008), despite lacking a brain with which to process visual input. The extent to which the eyes of cubozoans and bilaterians reflect convergent evolution is unclear; vision is mandatory to the cubozoan predatory lifestyle, and some form of photosensitivity is likely to have been an ancestral characteristic of the phylum Cnidaria. Although limited molecular data are as yet available, the hypothesis that vision evolved independently within Cnidaria cannot yet be rejected; it is quite possible that the eyes of cnidarians and bilaterians reflect independent elaborations on nervous systems that have common origins.

Whereas there is general agreement that the nervous systems of cnidarians and bilaterians have a common origin, the situation in ctenophores is much less clear. Peptides are thought to be the predominant neurotransmitters employed in cnidarian nervous systems, but GABA, glutamate, serotonin, acetylcholine, and other classical neurotransmitters are almost certainly also used (Anctil 2009). The ctenophore nervous system, however, appears to be based exclusively on glutamate transmission—although GABA is present, this is likely to be a breakdown product, and there is no evidence for the use of classical neurotransmitters. Moreover, the limited representation of genes typically involved in bilaterian neurogenesis has been used to argue for independent origins of the ctenophore nervous system (Moroz et al. 2014).

Parsimony is usually the guiding principle in evolutionary biology, and this kind of thinking urges us to always think in terms of common origins—for nervous systems, eyes, and other complex structures. However, in the real world, evolution sometimes follows non-parsimonious paths. Vision confers an enormous selective advantage, as does a nervous system, so too with striated musculature permitting rapid movement. Counterintuitively, striated muscle has evolved independently in the Bilateria and the Cnidaria (Steinmetz et al. 2012), leading us to suggest that the assumption of parsimony itself deserves greater scrutiny. The temptation to embrace claims of common origins is great, and too frequently the possibility of convergence is not seriously considered.

3.7 The Closest Unicellular Relatives of Extant Animals

Complex multicellularity has arisen very few times during the long history of life on earth, and the Metazoa have been by far the most successful of these ventures, hence, there is enormous interest in how this transition came about. One approach to this complex issue is to compare the gene sets of animals with those of their closest unicellular relatives.

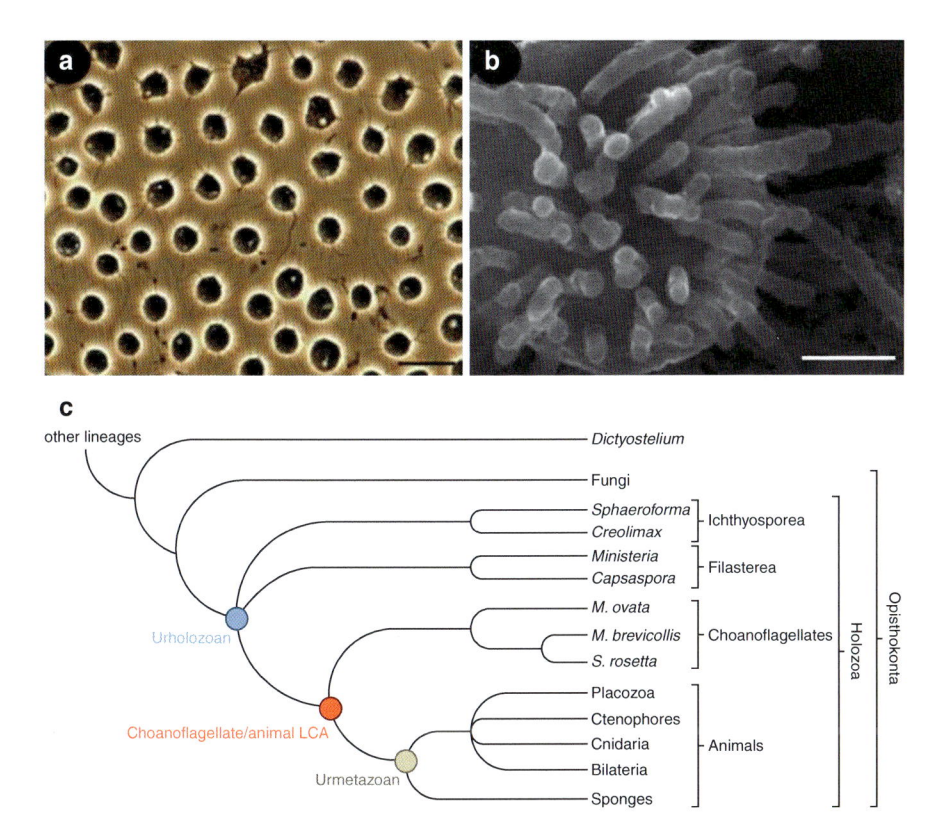

Fig. 3.12 The unicellular holozoan *Capsaspora* and its evolutionary position. (**a**) Differential interference contrast and (**b**) scanning electron microscopy images of *Capsaspora owczarzaki*. The scale bar represents 10 and 1 μm, respectively, in the two images (images from Suga et al. 2013). (**c**) Holozoan evolutionary relationships. The Holozoa consists of kingdom Metazoa and its unicellular relatives, the choanoflagellates, filastereans (which include *Capsaspora*), and ichthysporeans (From Richter and King 2013)

Animals belong to the greater clade Holozoa, which also contains several unicellular lineages (Fig. 3.12). The significance of one of these, the choanoflagellates, has been alluded to above on the basis of their striking similarity with the choanocytes of sponges. The choanoflagellates are the closest relatives of the Metazoa and are microscopic consumers of bacteria. Most choanoflagellates are unicellular, but species with colonial life cycle stages are also known. The close relationship between choanoflagellates and animals and the multicellularity of sorts seen in some representatives led to the hope that this group would enable deep insights into the origins of metazoan multicellularity, but this hope proved false. Elegant work from Nicole King and others (Alegado et al. 2012) demonstrated that in the case of the choanoflagellate *Salpingoeca rosetta*, formation of rosettes (the colonial stage) is triggered by a sulfolipid component of the cell wall of the prey bacterium. Hence choanoflagellate coloniality represents a distinct experiment in multicellularity that is not informative with respect to animal origins.

Nevertheless, genome sequencing of representative choanoflagellates (*Monosiga brevicollis, S. rosetta*) and other unicellular holozoan lineages (such as *Capsaspora owczarzaki*) is enabling a much better understanding of the gene complement of Urmetazoa—the common ancestor of all animals. It is now clear that far fewer genes distinguish metazoans than was assumed to be the case—cadherins, integrins, and many "animal-specific" transcription factors and signaling pathway components have been identified in non-metazoans (Suga et al. 2013).

3.8 The Paucity of Data on Symbioses Involving "Lower" Animals

Given the diversity of non-bilateral animals outlined above, the focus of the rest of this book on two groups of cnidarians may appear very narrow. However, given the preliminary nature of much of the data available for other "lower" animals, at this time attempts to broaden the scope would be premature. It is unquestionably true that, in terms of microbial interactions, the *Hydra* holobiont is the best understood representative of the four non-bilateral phyla discussed here, and it is clear that it will continue to provide insights into the mechanisms of animal–microbe interactions. It could be argued that the material included on coral–microbe interactions is somewhat premature given the state of this field. While this may be true, this is a dynamic field that attracts a great deal of public interest—coral disease and the future of coral reefs are major general concerns, and it is important that our understanding of coral–microbe interactions is substantially improved in short order, to enable more enlightened management of reef resources faced with ever-increasing anthropogenic challenges. For these reasons, we make some generalizations that history may dictate be revisited in the near future and hope that our readers will not judge us too harshly for having made them.

References

Alegado RA et al (2012) A bacterial sulfolipid triggers multicellular development in the closest relatives of animals. eLife 1:e00013

Anctil M (2009) Chemical transmission in the sea anemone Nematostella vectensis: a genomic perspective. Comp Biochem Physiol Part D Genomics Proteomics 4:268–289

Ball E, Hayward D, Saint R et al (2004) A simple plan – cnidarians and the origins of developmental mechanisms. Nat Genet 5:567–577

Ball E, DeJong D, Schierwater B et al (2007) Implications of cnidarian gene expression patterns for the origins of bilaterality - is the glass half full or half empty? Integrative and Comparative Biology 47(5):701–711

Chapman JA et al (2010) The dynamic genome of Hydra. Nature 464:592–596

Dayel MJ, Alegado RA, Fairclough SR, Levin TC, Nichols SA, McDonald K, King N (2011) Cell differentiation and morphogenesis in the colony-forming choanoflagellate Salpingoeca rosetta. Dev Biol 357(1):73–82.

Eitel M et al (2013) Global diversity of the Placozoa. PLoS One 8:e57131

Jager M, Dayraud C, Mialot A, Quéinnec E, le Guyader H, et al (2013) Evidence for involvement of Wnt signalling in body polarities, cell proliferation, and the neuro-sensory system in an adult ctenophore. PLoS ONE 8(12):e84363

King N (2004) The unicellular ancestry of animal development. Dev Cell 7:313–325

King N (2005) Choanoflagellates. Curr Biol 15(4):R113–R114

Kortschak RD et al (2003) EST analysis of the cnidarian Acropora millepora reveals extensive gene loss and rapid sequence divergence in the model invertebrates. Current Biology 13: 2190–2195

Kozmik Z et al (2008) Assembly of the cnidarian camera-type eye from vertebrate-like components. Proc Natl Acad Sci U S A 105:8989–8993

Miller DJ, Ball EE (2008) Cryptic complexity captured: the Nematostella genome reveals its secrets. Trends Genet 24:1–4

Moroz LL et al (2014) The ctenophore genome and the evolutionary origins of neural systems. Nature 510:109–114

Putnam NH, Srivastava M, Hellsten U, Dirks B, Chapman J, Salamov A, et al (2007). Sea anemone genome reveals ancestral eumetazoan gene repertoire and genomic organization. Science 317:86–94

Richter DJ, King N (2013) The genomic and cellular foundations of animal origins. Ann Rev Genet 47:509–537

Shinzato C et al (2011) Using the Acropora digitifera genome to understand coral responses to environmental change. Nature 476:320–323

Siegl A et al (2011) Single-cell genomics reveals the lifestyle of Poribacteria, a candidate phylum symbiotically associated with marine sponges. ISME J 5:61–70

Smith CL et al (2014) Novel cell types, neurosecretory cells and body plan of the early-diverging metazoan Trichoplax adhaerens. Curr Biol 24:1565–1572

Srivastava M et al (2008) The Trichoplax genome and the nature of placozoans. Nature 454:955–960

Steinmetz PR et al (2012) Independent evolution of striated muscles in cnidarians and bilaterians. Nature 487:231–234

Suga H et al (2013) The Capsaspora genome reveals a complex unicellular prehistory of animals. Nat Commun 4:2325

Tang F, Bengtson S, Wang Y, Wang XL, Yin CY (2011) Eoandromeda and the origin of Ctenophora. Evol Dev 13(5):408–14

Technau U et al (2005) Maintenance of ancestral genetic complexity and non-metazoan genes in two basal cnidarians. Trends Genet 21:633–639

Valentine JW, Jablonski D, Erwin DH (1999) Fossils, molecules and embryos: new perspectives on the Cambrian explosion. Development 126:851–859

Williams GC (1994) Biotic diversity, biogeography and phylogeny of pennatulacean octocorals associated with coral reefs on the Indo-Pacific. Proc Seventh Int Coral Reef Symp 2:729–735

Williams GC (2011) The global diversity of sea pens (Cnidaria: Octocorallia: Pennatulacea). PLoS One 6, e22747

Xiao X, Laflamme M (2009) On the eve of animal radiation: phylogeny, ecology and evolution of the Ediacara biota. Trends Ecol Evol 24:31–40

Phylosymbiosis: Novel Genomic Approaches Discover the Holobiont

4

Important questions in biology are usually raised long time ago. Answering them, however, has to await the development of appropriate techniques. Due to the emergence and rapid technological advances in culture-independent techniques to identify and characterize microbes, particularly genomic approaches, we learned that organisms from *Hydra* to man are to be considered holobionts stably associated with bacteria, mainly strict anaerobes, but also including viruses, archaea, fungi, and protists.

The increasing realization that animals cannot be considered in isolation but only as a partnership of animals and symbionts has led to two important realizations. First, it is becoming increasingly clear that to understand the physiology, evolution, and development of a given species, we cannot study the species in isolation. And second, the health and fitness of animals, including humans, appears to be fundamental multiorganismal. Any disturbance within the complex community has drastic consequences for the well-being of the members. And third, the holobiont may be an important unit of evolutionary selection, a selection of "teams" containing many genomes and species.

4.1 Animal Life and Fitness Is Fundamental Multiorganismal

First insights into the microbiota within us came for the mouse and human gut microbiome. There, huge efforts by large-scale metagenomic studies such as the National Institutes of Health-funded Human-Microbiome Project (http://commonfund.nih.gov/hmp), the European-funded Metagenomics of the Human Intestinal Tract (http://metahit.eu) consortium, and the International Human Microbiome Consortium (http://www.human-microbiome.org) have helped to determine the breadth of microbial variation and function across large populations, revealed a high interindividual diversity and stability over time and established associations between microbiota alterations and disease states (Weinstock 2012; Morgan et al. 2013;

© Springer-Verlag Wien 2016
T.C.G. Bosch, D.J. Miller, *The Holobiont Imperative: Perspectives from Early Emerging Animals*, DOI 10.1007/978-3-7091-1896-2_4

Human Microbiome Consortium, Nature 2012a, b). Mostly due to the efforts of the Jeff Gordon and Robert Knight laboratories, we learned that the adult human gut is inhabited by 10^{13} to 10^{14} microorganisms, a figure thought to be at least 10 times greater than the number of human cells in our bodies with 150 times as many genes as our genome (Lozupone et al. 2012). The estimated species number varies greatly, but it is generally accepted that the human microbiome consists of greater than 1000 species and more than 7000 strains.

4.2 Phylosymbiosis and Coevolution

Eukaryotic evolution has probably never seen a period without the presence of microbes. There is no germ-free animal around. Appreciation grows for the fact that animals live in a bacterial world (McFall-Ngai et al. 2013), and no longer can the animal be viewed as separate from the microbes it requires to subsist, reproduce, and evolve over time. Microbiota colonization can depend on the genetics of the host, and there is an intensifying interest today in resolving the relative contributions of the environment and host genes on the assembly of host-associated microbial communities. The majority of microbial entities in eukaryotes are not transient passengers or tourists randomly acquired from the environment, but instead function with specific roles in eukaryotic nutrition, reproduction, development, and maybe even speciation. Thus, symbiosis is a major component of eukaryotic fitness and evolution.

Relatively little is known about the comparative structure and evolution of bacterial communities among closely related host species, but this knowledge gap is starting to change. In a seminal paper, Brucker and Bordenstein (2012a) hypothesized that if host phylogenetic relationships, in part, structure gut microbial communities, then related species of animals reared on the same diet will not acquire the same microbiome, but instead host species-specific communities of microbes.

Seth Bordenstein at Vanderbit University subsequently introduced the term "phylosymbiosis" to describe the history of changes in the total microbial communities of related host species (Fig. 4.1) Brucker and Bordenstein (2013a, b). Phylosymbiotic is akin to phylogenomic, but the focus is on inferring the relationships of total microbial assemblages across related host species, rather than the relationships of host genes. Differential microbial compositions occur between closely related species when maintained on the same diet and under identical rearing conditions, and the community relationships of each species' microbiome parallel the phylogenetic relationships of the host genome. Thus, the term "phylosymbiosis" describes the outcome of such relationships as an evolutionary signal of host phylogeny and microbial community co-structuring between species. Does this mean that all species-specific communities of microbes must diverge in parallel with the host during the formation of a new host species? Can, therefore, the relationships between microbial communities predict the evolutionary relationships of the host species and vice versa? Do phylosymbiotic relationships always indicate reciprocal evolutionary change between the groups of interacting species—and thereby coevolution?

4.2 Phylosymbiosis and Coevolution

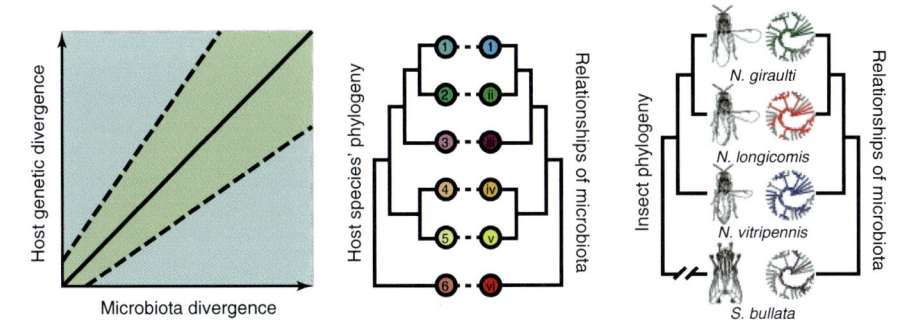

Fig. 4.1 Phylosymbiosis. Divergence in host genes is positively correlated with the differentiation of the microbiome. Like phylogenomics, phylosymbiosis is a total microbiome metric that retains an ancestral signal of the host's evolution. (*left*) The central prediction is that divergence in host genes is positively correlated with differentiation of the microbiome. (*middle*) Parallel dendrogram between the host phylogeny and the microbiome relationships is one test of phylosymbiosis. (*right*) Schematic of a real data example from a model study in insects (Taken from Brucker and Bordenstein 2012a)

Here, we enter controversial ground. Evolutionary biologists have a strong opinion about the term "coevolution." The term often is used to describe cases where two or more species reciprocally affect each other's evolution. Coevolution is likely to happen when different species have close ecological interactions with one another. So, for example, an evolutionary change in the morphology of a plant might affect the morphology of an herbivore that eats the plant, which in turn might affect the evolution of the plant, which might affect the evolution of the herbivore…and so on Futuyma (2005). At first glance, it seems that everything is involved in coevolution since virtually all organisms interact with other organisms and presumably influence their evolution in some way. But this assumption depends entirely on one's definition of the term "coevolution." Coevolution is defined by Futuyma (1979, p. 453; see also Ehrlich and Raven, 1965) as "… *evolution in which fitness of each genotype depends on the population densities and genetic composition of the species itself and the species with which it interacts.*" Futuyma and Slatkin (1983) further comment "… *the study of coevolution is the analysis of reciprocal genetic changes that might be expected to occur in two or more ecologically interacting species and the analysis of whether the expected changes are actually realized*". Thus, like the issue of defining an adaptation, we should not invoke coevolution without reasonable evidence that the traits in each species were a result of or evolved from the interaction between the two species. A very general observation of one group of organisms having an influence on another group of organisms is not enough to indicate coevolution as long as the reciprocal changes have not been documented clearly. And any evolutionary old interaction, symbiosis, mutualism, etc., are not synonymous with coevolution. In one sense, there has definitely been "evolution together" but whether this fits the strict definition of coevolution needs to be determined by careful observation, experimentation, and phylogenetic analysis. In particular, phylogenetic analysis is a crucial component of coevolution. If the cladograms of the host and the cladograms of the symbiont are congruent (see Figure 4.1), this certainly suggests coevolutionary

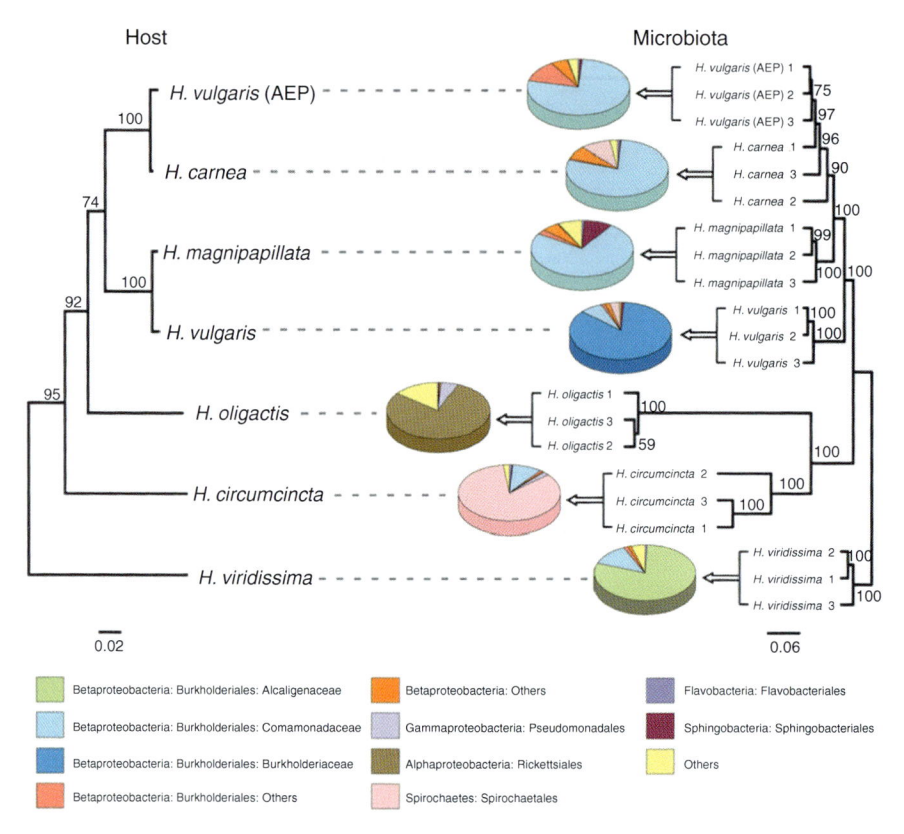

Fig. 4.2 Hydra species are colonized by species-specific microbiota. Composition of the microbiome in *Hydra* parallels the phylogenetic relationships of the *Hydra* species (Taken from Franzenburg et al. 2013)

phenomena. With regard to our discussion on phylosymbiosis, we hope to have made it clear that one needs to know the evolutionary history before one can make firm statements about "coevolution." Let us turn to the early emerging metazoans and to a textbook example of phylosymbiosis. In 2007, a graduate student with some training in microbiology, Sebastian Fraune, joined the Bosch lab to look into the microbiome of two widely used species of the freshwater polyp *Hydra*, *H. vulgaris,* and *H. oligactis*. The young researcher soon uncovered that two species of the cnidarian *Hydra* are colonized by remarkably different bacterial communities, although being cultured under identical laboratory conditions and same artificial diet (brine shrimp nauplii) for decades. Even more astonishing, these laboratory cultures were colonized by microbial communities which were found to be very similar to that of the same *Hydra* species freshly isolated from lakes in Northern Germany. These observations could only mean that there are strong host-mediated selective forces on the associated microbiota (Fraune and Bosch 2007). In a subsequent study, we extended these observations to many more *Hydra* species and were able to show phylosymbiotic host–microbe relationships in seven species. The studied species were separately reared in simple, water-containing plastic dished for up to 30 years under identical environmental conditions, including standardized diet. Nevertheless, their microbial composition differed substantially and revealed a highly phylosymbiotic pattern (Fig. 4.2). Impressively, after three decades of identical cultivation, each species still maintained its specific, bacterial fingerprint. *Hydra*'s microbiome, therefore, certainly reflects an ancestral footprint of evolution.

Searching for the mechanisms behind this remarkable specificity, a subsequent study from the Bosch lab uncovered that important components of *Hydra*'s innate immune system, antimicrobial peptides, play a major role. An elaborated experimental setup including both transgenic polyps and germ-free ("axenic") polyps in common garden experiments uncovered a group of species-specific antimicrobial peptides called arminins as critical determinants of the microbiota assembly. Common garden experiments can provide a more mechanistic understanding of the causes of compositional and spatial variation in host–microbe interactions. When axenic *Hydra* polyps in a culture dish were exposed to polyps with bacterial communities characteristic for different *Hydra* species, the recipient host selected for bacterial taxa resembling its native microbiota, just like in an elegant reciprocal microbiota, transfer experiments between zebrafish and mice conducted by Rawls et al. (2006). However, arminin loss-of-function polyps significantly lost this selective potential and ended up with untypical bacterial communities. When inoculated with their native bacterial colonizers, control- and arminin-deficient *Hydra* were colonized equally, indicating that the species-specific microbiota is partially resistant to the antimicrobial peptides expressed by its host. Thus, as discussed in chapter six, the host's immune system indeed plays a major role in selecting the bacterial associates. How do animals change their microbial partners? The abovementioned Franzenburg et al. paper (2013) suggests that changes in fast evolving antimicrobial peptides are sufficient to drastically alter host-associated bacterial communities. Coupling fast evolving genes with the adaptive potential of changed bacterial partners could be a potential promoter for speciation.

Since we are able to see that the phylogeny of the *Hydra* host completely mirrors that of the microbiome (Fraune and Bosch, 2007; Franzenburg et al. 2013), the concept of "phylosymbiosis" appears well documented in the cnidarian *Hydra*. Similar observations were made in social insect by Brucker and Bordenstein. They discovered in *Nasonia* wasps that the gut microbiome was different between closely related species of insects reared on the same diet, and the constituents and composition of the bacterial communities in each species changed in parallel with the genomic relationships of the host species (Brucker and Bordenstein 2012a, b). The enduring association of a given microbiome with the *Hydra* or *Nasonia* or mammalian host is astonishing. How old are such associations? How old is a given *Hydra* species?

4.3 Microbiota Diversification Within a Phylogenetic Framework of Hosts: Insights from *Hydra*

Despite *Hydra*'s long history as a model organism for animal evolution, key features of the evolution of *Hydra* itself such as the evolutionary origins of *Hydra* and its species level diversity are still not well understood. As a consequence, the pace and timeframe of crucial evolutionary processes and novelties, like the emergence of numerous orphan genes specific for *Hydra* or some of its species (Khalturin et al. 2008, 2009), cannot be assessed Cartwright and Collins (2007). Moreover, current lack of information on precise phylogenetic placement of the Hydra species, however, makes it impossible to investigate the diversity of the *Hydra*-specific microbiota within a coevolutionary framework. Resolving the evolutionary origins of *Hydra* has been impeded by the absence of fossilized remains. Recent phylogenetic analyses unambiguously placed *Hydra* within the hydrozoan taxon Aplanulata, but the age of *Hydra* and the timing of its diversifications remained largely unknown. Different molecular clock approaches diverged greatly in their estimates: while the age of the viridissima group was estimated to be 156–174 million years based on the divergence of its symbiotic Chlorella (Kawaida et al. 2013), the age of crown group *Hydra* was estimated to be only ~60 million years based on assumed substitution rates for COI and 16S (Martínez et al. 2010; Nawrocki et al. 2013).

To get a rather precise idea about the age of *Hydra* and its species, we used known Cnidarian fossils (e.g., Chen et al. 2002; Love et al. 2009; Waggoner and Collins 2004; Xiao et al. 2000; Young and Hagadorn 2010) as calibration points to infer divergence times among *Hydra* lineages and accounted for rate heterogeneity among lineages by applying relaxed-clock models (Schwentner and Bosch 2015). Our divergence date estimates indicate for the crown group *Hydra* an age of 193 million years ago (mya), representing the split between the *viridissima* group and all other *Hydra*. The inferred sister taxon (*Candelabrum cocksii*) diverged 326 mya. Within the *Hydra viridissima* group has an estimated age of 112 mya, the braueri group of 69 mya, the oligactis group of 71 mya, and the vulgaris group of 62 mya. The latter two groups diverged about 101 mya and in turn separated from the braueri group 147 mya. The divergence time estimates obtained date the evolutionary origins of crown group *Hydra* to the late Triassic or early Jurassic (Fig. 4.3). Since in this period, all continental plates were still part of Pangaea, today's global distribution of Hydra could be a result of vicariance with little or no large-scale (i.e., without transoceanic) dispersal. But the divergence between the *oligactis*, *braueri*, and *vulgaris* groups and the ages of each of these groups are younger than the separation between Gondwana and Laurasia. Even Gondwana most likely broke apart before the four extant groups of *Hydra* evolved (Upchurch 2008). Consequently, the cosmopolitan distributions of the vulgaris group and most likely also of the *viridissima* group are due to transcontinental and transoceanic dispersal. It is noteworthy that the diversification of three *Hydra* groups—*oligactis*, *braueri*, and *vulgaris* groups—as well as the diversification within the two largest monophyletic clades within the *viridissima* group occurred between ~60 and ~70 mya (Fig. 4.4). This coincides roughly with the Cretaceous–Paleogene boundary, a time of severe climatic change and of global mass extinction. Because the majority of extant species occur in North America and/or Eurasia, the ancestors of today's groups most likely survived on Laurasia and spread further south from there. Transoceanic dispersal after the separation of all main land masses would explain why *Hydra* species from Africa,

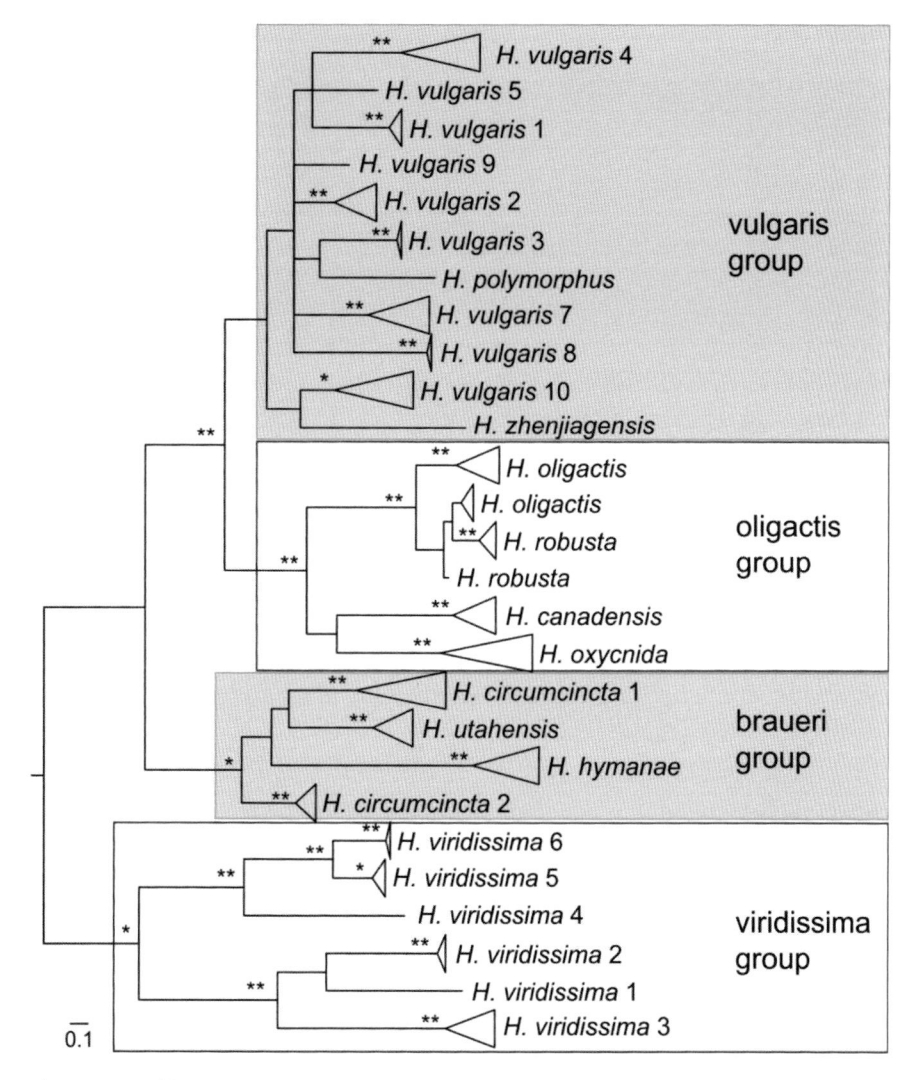

4.3 Microbiota Diversification Within a Phylogenetic Framework of Hosts

Fig. 4.3 Identification of "main lineages" (= hypothetical species) of *Hydra* by independently analyzing three molecular markers: mitochondrial COI (cytochrome c oxidase subunit I), mitochondrial 16S, and nuclear ITS region (internal transcribed spacer; spanning ITS1, 5.8S, and ITS2) (From Schwentner and Bosch 2015)

South America, and Australia do not show the close relationship expected for taxa with Gondwana distributions; however, this may also be due to the overall low resolution of interspecific relationships.

Our divergence time estimates set the time frame not only to date the origins of evolutionary novelties but also to assess the rates of genomic changes for *Hydra* and its groups and species to accommodate defined bacterial associates that are essential

Fig. 4.4 The diversification of *Hydra* groups—occurred between ~60 and ~70 mya. This coincides with the Cretaceous–Paleogene boundary, a time of severe climatic change and of global mass extinction

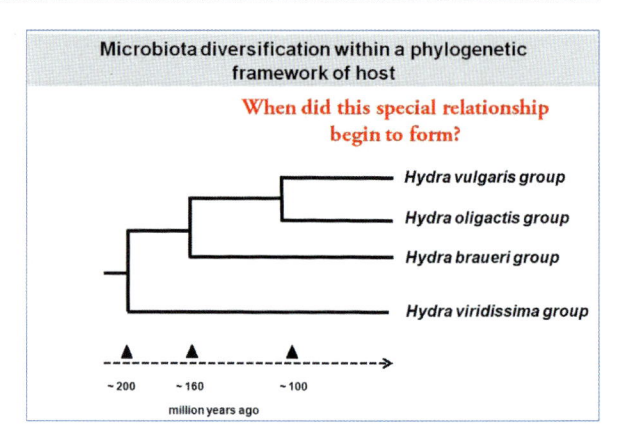

for the host's fitness. If this ability is genetically fixed in the host's genome, or if the host's genotype matters for microbiota composition, closer related host species should be colonized by more similar bacterial communities compared to distantly related species. Future experiments will allow to assess the role of microbes in reproductive isolation, and hence speciation, as well as explore the evolutionary consequences of phylosymbiosis in these early emerging animals. The piece of the puzzle that next needs our attention is how the individual members of the holobiont talk to each other.

References

Brucker RM, Bordenstein SR (2012a) Speciation by symbiosis. Trends Ecol Evol 27(8):443–451
Brucker RM, Bordenstein SR (2012b) The roles of host evolutionary relationships (genus: Nasonia) and development in structuring microbial communities. Evolution 66(2):349–362
Brucker RM, Bordenstein SR (2013a) The capacious hologenome. Zoology 116(5):260–261
Brucker RM, Bordenstein SR (2013b) The hologenomic basis of speciation: gut bacteria cause hybrid lethality in the genus Nasonia. Science 341(6146):667–669
Cartwright P, Collins A (2007) Fossils and phylogenies: integrating multiple lines of evidence to investigate the origin of early major metazoan lineages. Integr Comp Biol 47:744–751
Chen J-Y, Oliveri P, Gao F, Dornbos SQ, Li C-W, Bottjer DJ, Davidson EH (2002) Precambrian animal life: Probable developmental and adult cnidarian forms from southwest China. Dev Biol 248:182–196
Fraune S, Bosch TCG (2007) Long-term maintenance of species-specific bacterial microbiota in the basal metazoan Hydra. Proc Natl Acad Sci U S A 104:13146–13151
Franzenburg S, Fraune S, Altrock PM, Künzel S, Baines JF, Traulsen A, Bosch TCG (2013) Bacterial colonization of Hydra hatchlings follows a robust temporal pattern. ISME J 7(4): 781–90
Futuyma DJ (1979) Evolutionary biology, 1st edn. Sinauer Associates, Sunderland. ISBN 0-87893-199-6
Futuyma DJ (2005) Evolution. Sinauer Associates, Sunderland. ISBN 0-87893-187-2
Futuyma DJ, Slatkin M (eds) (1983) Coevolution. Sinauer Associates, Sunderland. ISBN 0-87893-228-3

Kawaida H, Ohba K, Koutake Y, Shimizu H, Tachida H, Kobayakawa Y (2013) Symbiosis between Hydra and Chlorella: molecular phylogenetic analysis and experimental study provide insight into its origin and evolution. Mol Phylogenet Evol 66:906–914

Khalturin K, Anton-Erxleben F, Sassmann S, Wittlieb J, Hemmrich G, Bosch TCG (2008) A novel gene family controls species-specific morphological traits in Hydra. PLoS Biol 6:e278

Khalturin K, Hemmrich G, Fraune S, Augustin R, Bosch TCG (2009) More than just orphans: are taxonomically-restricted genes important in evolution? Trends Genet 25:404–413

Love GD, Grosjean E, Stalvies C, Fike DA, Grotzinger JP, Bradley AS, Kelly AE, Bhatia M, Meredith W, Snape CE, Bowring SA, Condon DJ, Summons RE (2009) Fossil steroids record the appearance of Demospongiae during the Cryogenian period. Nature 457:718–721

Lozupone CA, Stombaugh JI, Gordon JI, Jansson JK, Knight R (2012) Diversity, stability and resilience of the human gut microbiota. Nature 489:220–230

Martínez DE, Iñiguez AR, Percell KM, Willner JB, Signorovitch J, Campbell RD (2010) Phylogeny and biogeography of Hydra (Cnidaria: Hydridae) using mitochondrial and nuclear DNA sequences. Mol Phylogenet Evol 57:403–410

McFall-Ngai M, Hadfield M, Bosch T, Carey H, Domazet-Loso T, Douglas A, Dubilier N, Eberl G, Fukami T, Gilbert S, Hentschel U, King N, Kjelleberg S, Knoll A, Kremer N, Mazmanian S, Metcalf J, Nealson K, Pierce N, Rawls J, Reid A, Ruby E, Rumpho M, Sanders J, Tautz D, Wernegreen J (2013) Animals in a bacterial world, a new imperative for the life sciences. Proc Natl Acad Sci U S A 110(9):3229–3236

Morgan XC, Segata N, Huttenhower C (2013) Biodiversity and functional genomics in the human microbiome. Trends Genet 29(1):51–58

Nawrocki AM, Collins AG, Hirano YM, Schuchert P, Cartwright P (2013) Phylogenetic placement of Hydra and relationships within Aplanulata (Cnidaria: Hydrozoa). Mol Phylogenet Evol 67:60–71

Rawls JF, Mahowald MA, Ley RE, Gordon JI (2006) Reciprocal gut microbiota transplants from zebrafish and mice to germ-free recipients reveal host habitat selection. Cell 127(2):423–433

Schwentner M, Bosch TCG (2015) Revisiting the age, evolutionary history and species level diversity of the genus Hydra (Cnidaria: Hydrozoa). Mol Phylogenet Evol 91:41–55

The Human Micorbiome Project Consortium (2012a) A framework for human microbiome research. Nature 486:215–221

The Human Micorbiome Project Consortium (2012b) Structure, function and diversity of the healthy human microbiome. Nature 486:207–214

Upchurch P (2008) Gondwanan break-up: legacies of a lost world? Trends Ecol Evol 23:229–236

Waggoner B, Collins AG (2004) Reducto ad absurdum: testing the evolutionary relationships of ediacaran and paleaozoic problematic fossils using molecular divergence dates. J Paleont 78:51–61

Weinstock GM (2012) Genomic approaches to studying the human microbiota. Nature 489:250–256

Xiao S, Yuan X, Knoll AH (2000) Eumetazoan fossils in terminal Proterozoic phosphorites? Proc Natl Acad Sci U S A 97:13684–13689

Young GA, Hagadorn JW (2010) The fossil record of cnidarian medusae. Palaeoworld 19:212–221

Negotiations Between Early Evolving Animals and Symbionts

5.1 Cnidaria Use a Variety of Molecular Pathways to Elicit Complex Immune Responses

Cnidarians are constantly exposed to microbes. It is therefore not too surprising that molecular analyses have revealed a variety of molecular pathways used by cnidarians to respond to microbial exposure. When approaching the epithelium of a cnidarian, microbes first encounter extracellular germline-encoded surface receptors that recognize microbe-associated molecular patterns (MAMPs). Examples of MAMPs include lipopolysaccharides (LPSs), peptidoglycan (PGN), flagellin, and microbial nucleic acids. Perception of a MAMP at the cell surface is achieved by Toll-like receptors (TLRs) which then initiate MAMP-triggered immunity. In this chapter, we consider one by one the major components used by the cnidarians host to interact with the microbes.

Our most complete understanding of the cnidarian response to MAMPs relates to perception via the Toll/Toll-like receptor (TLR) pathway (Augustin et al. 2010; Bosch 2013). Toll-like receptors (TLRs) are conserved throughout animal evolution. A canonical Toll/TLR pathway is present in representatives of the basal cnidarian class Anthozoa, but neither a classic Toll/TLR receptor nor a conventional nuclear factor (NF)-κB could be identified in the hydrozoan *Hydra*. There, two genes could be identified whose inferred amino acid sequence contained a Toll/interleukin-1 receptor (TIR) domain, a transmembrane domain, and an extracellular domain lacking any specific domain structure. We have termed these *Hydra* genes Toll-receptor-related 1 (HyTRR-1) and Toll receptor- related 2 (HyTRR-2) respectively. Most strikingly, neither HyTRR-1 nor HyTRR-2 contains leucine-rich repeats (LRRs) in its extracellular region. TLR function in *Hydra* (Fig. 5.1) is realized by the interaction of a leucine-rich repeat (LRR) domain containing protein with a Toll/interleukin-1 receptor (TIR)-domain-containing protein lacking LRRs. This receptor complex activation then triggers the innate immune response which involves the production of species-specific antimicrobial peptides such as periculin (Bosch et al. 2009; for more references see "suggested reading").

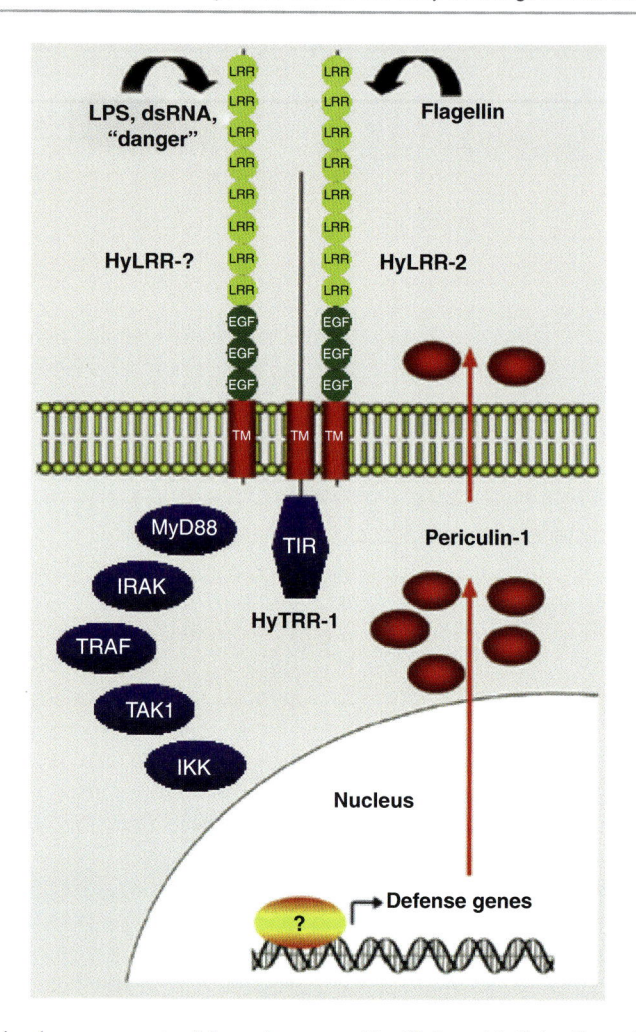

Fig. 5.1 Molecular components of the pathways used by *Hydra* epithelial cells to elicit complex immune responses. *TLR* Toll-like receptors, *LRR* leucine-rich repeat domain, *LPB* lipopolysaccharide-binding protein, *CD14* pattern recognition protein, *TIR* Toll/IL-1R domain, *MyD88* Myeloid differentiation primary response gene, *IRAK* IL-1R-associated kinase family, *TRAF* TNF receptor-associated factor protein family, *TAK1* transforming growth factor β (TGF-β)-activated kinase 1, *IKK* inhibitor of nuclear factor kappa-B kinase, *NF-κB* nuclear factor kappa-light-chain-enhancer of activated B cells, *AMP* antimicrobial peptide

Microbes evading membrane-bound TLR receptors or specifically invading epithelial cells encounter another line of defense inside the cnidarians host cell, the NOD-like receptors (NLRs). Homologs of the NLRs (e.g., R genes) have been discovered throughout the plant and animal kingdoms, including cnidarians, the sea urchin and the zebra fish. The high evolutionary conservation of the NLRs underlines their significance in host defense. Similar to sea urchin which has at least 203 identified putative NLRs, cnidarians such as *Nematostella* and *Hydra* have hundreds of NLRs. In a systematic survey of the NACHT and NB-ARC domain genes in existing expressed sequence tag (EST) and genome data sets, we found that the *Hydra* genome has 290 putative NBD loci falling into two large groups: 130 of the NACHT domain type and 160 of the NB-ARC domain type (Lange et al. 2010). The *Hydra* and Cnidaria NACHT/NB-ARC complements include novel combinations of domains. Thus, as in vertebrates, a broad repertoire of NLRs seems to be involved in the recognition of conserved microbial components in *Hydra*. While these observations clearly show that NLRs are ancient genes, their immune function and significance in host–microbe interactions remains to be determined. The main localization of NLR expression in *Hydra* is confined to the endodermal layer, where NLRs could play a crucial role in maintaining a balance with the commensals/symbionts of the gastric cavity. The advent of novel sequencing techniques will allow for the first time the analysis of the diversification of this class of genes in organisms at the base of animal evolution and will contribute to understand evolutionary pressures that have shaped genetic diversity profiles of immune genes.

Following recognition of conserved microbial features (MAMPS) by cell surface receptors, innate immune reactions in cnidarians are initiated from the cytoplasmic TIR domains of TLRs. For further signal transduction, a conserved adaptor, MyD88, has been identified in *Nematostella* as well as in *Hydra* which appears an essential component for the activation of innate immunity. MyD88 possesses the TIR domain in the C-terminal portion and a death domain in the N-terminal portion. MyD88 associates with the TIR domain of TLRs. Upon stimulation, MyD88 recruits IL-1 receptor-associated serine/threonine kinase (IRAK) to TLRs through interaction of the death domains of both molecules. IRAK is activated by phosphorylation and then associates with TRAF6, leading to the activation of two distinct signaling pathways and finally to the activation of JNK and NF-κB.

To address the function of MyD88 directly, a MyD88 loss-of-function study in *Hydra vulgaris* (AEP) using a stable transgenic *Hydra* line with reduced expression level of MyD88 was combined with microarray analyses to identify effector gene downstream of the TLR-signaling cascade. In parallel, the gene expression profile of germfree animals was examined to directly investigate the connection between TLR-signaling and bacterial recognition. More than 75 % of the MyD88-responsive transcripts appeared to be also altered in germfree polyps, indicating that expression of these genes is responsive to bacterial recognition. These observations show that recognition of bacteria is an ancestral function of TLR signaling, which is important for the selection and maintenance of resident microbiota, and contributes to defense against bacterial pathogens (Bosch 2013).

Downstream of the conserved signaling cascade are conserved stress-responsive transcription factors including structurally related eukaryotic transcription factors of the NF-kappaB (NF-kB) family of proteins. The activity of NF-kB is primarily regulated by interaction with inhibitory IkB proteins. In most cells, NF-kB is present as a latent, inactive, IkB-bound complex in the cytoplasm. When a cell receives any of a multitude of extracellular signals, NF-kB rapidly enters the nucleus and activates the expression of a rich repertoire of mostly not conserved target genes.

5.2 How Do Cnidarians Distinguish Between Friends and Foes: Insights from Corals and *Hydra*

Cnidarians lack conventional antibody-based immunity but harbor large numbers of beneficial microorganisms. How are pathogens eliminated and long-lived and specific symbiotic associations established and maintained? There is increasing evidence that the inter-partner signaling pathways involved during the onset of symbiosis are homologous to those driving animal host/pathogenic microbe interactions. In both types of relationships, there are common principles, and besides released signals, pathogen/symbiont surface molecules (PAMPs/MAMPs) most likely are major determinants for an interaction with the host. The cellular and molecular interactions underlying this interaction with particular emphasis on the establishment, maintenance, and breakdown of these cooperative partnerships are currently investigated in several coral species, including the Hawaiian stony coral *Fungia scutaria*; the tropical sea anemone *Aiptasia pallida*; a temperate sea anemone found on the Oregon coast, *Anthopleura elegantissima*; and a Red Sea soft coral, *Heteroxenia fuscescens*. Most species must acquire symbionts anew with each generation and therefore must engage in a complex recognition and specificity process that results in the establishment of a stable symbiosis.

To identify genes that initiate, regulate, and maintain this host/symbiont interaction, the Weis lab at Oregon State University has conducted a comparative transcriptome analysis in the host sea anemone *Anthopleura elegantissima* using a cDNA microarray platform. Although statistically significant differences in host gene expression profiles were detected between *Anthopleura elegantissima* in a symbiotic and nonsymbiotic state, the group of genes whose expression is altered is diverse, suggesting that the molecular regulation of the symbiosis is governed by changes in multiple cellular processes. To search for symbiosis-specific proteins during the natural onset of symbiosis in early host ontogeny, Barneah et al. (2006) used 2-dimensional polyacrylamide gel electrophoresis and compared patterns of proteins synthesized in symbiotic and aposymbiotic primary polyps of the Red Sea soft coral *Heteroxenia fuscescens* in the initiation phase, in which the partners interact for the first time. Surprisingly there were no changes detectable in the host proteome as a function of symbiotic state. To examine molecular mechanisms during initiation and establishment of coral–algae symbioses, Medina and colleagues followed gene expression profiles of coral larvae of *Acropora palmata* and *Montastraea faveolata* after exposure to *Symbiodinium* strains that differed in their ability to establish symbioses. Most interestingly, the coral host transcriptome remained almost unchanged during infection by symbionts that were able to establish symbioses. It was, however, massively altered during infection by symbionts that were not able to establish symbioses. Taken together, both the transcriptome and proteomics study do not support the existence of symbiosis-specific host genes involved in controlling and regulating the symbiosis. Instead, it appears that regulation of the immune system, apoptosis, and proteolysis may be the key players in the regulation of early coral–algae symbioses. "Successful coral–algal symbioses depend mainly on the symbionts' ability to enter the host in a stealth manner" (cited from Voolstra et al. 2009). If this scenario turns out to be true then several important questions await to be answered: How does the symbiont manage to survive within the host cell? Which molecules are released by the symbiont that participate in host–symbiont recognition and modification of the immune response? Which host pathways are responding to the molecules released by the symbiont? How is the delicate balance between successful metabolism and replication of the symbiont and survival of the host maintained on a long-term basis?

Observations in *Hydra* may teach us another lesson. *Hydra* can mount a powerful immune response in the absence of motile phagocytic cells. This response is contingent on two independent components: the presence of bacterial flagellin (a classic example of a MAMP) and the secretion of a tissue damage or danger signal, known as a damage-associated molecular pattern (DAMP). As outlined above, in *Hydra* two proteins are required for MAMP recognition and signal transduction: a *Hydra* Toll–interleukin-1 receptor domain (TIR domain)-containing protein (HyTRR) and *Hydra* leucine-rich-repeat protein 2 (HyLRR-2), containing a transmembrane domain. HyLRR-2 interacts with the HyTRR protein in response to flagellin. When exposed to a DAMP, such as monosodium urate (which is released from injured cells), an even stronger induction of defense genes is observed. For example, there is increased production of AMPs such as periculin-1. Nyholm and Graf proposed (2012) that the requirement for both a MAMP and a DAMP to launch an effective immune response constitutes a mechanism that allows *Hydra* to distinguish pathogens from symbionts. Further studies will show whether in *Hydra* this ability to integrate the two signals (MAMPs and DAMPs) may indeed contribute to shaping of the microbiota. In line with the conclusions drawn from coral studies, and based on the fact that AMPs such as periculins and arminins are instrumental in establishing the symbiotic bacterial community in *Hydra*, the innate immune system seems crucial for both the establishment and the long-term maintenance of host-associated bacterial communities in cnidarians.

5.3 Selection Can Favor the Establishment of Mutualisms and Animal–Microbe Cooperation

Setting aside the issue of how a close association between an animal and a microbe is first established, a requirement for this to become a stable relationship is mutual benefit—selection in favor of mutual dependency or "cooperation." If the microorganism imposes too high a cost on the host (i.e., becomes parasitic), then selection will favor evolution of efficient mechanisms for its recognition and elimination by the host. At the other end of the spectrum, the host may "learn" to exploit the microbe so efficiently that the latter is unable to exist independently, as has happened in the deep past in the case of mitochondria (which were derived from a *Rickettsia*-related bacterium) and much more recently in the strain of *Chlorella* that is responsible for the "green-ness" of the green hydra, *Hydra viridissima*. What we see in mutualistic symbioses may actually reflect the coincident overlap of "local minima" on the selection curves for the two partners—a semi-stable state (a "treaty" or "truce") which can be tipped in either direction by one partner evolving a new weapon.

In discussing the origins of multicellularity, theorists talk in terms of a period of "negotiation" between cells, which is perhaps also a useful concept in the context of understanding how mutualisms are established. Somehow or other, a microbe comes into intimate contact with an animal. If the costs on both sides are minimal, then the relationship will be tolerated (but not stable). If then small benefits occur to the host (e.g., a mutation leads to the microbe leaking a metabolite that is useful to the host), selection will favor the host evolving the means to recognize and "encourage" the microbe, perhaps by offering its own enticements ("you can have some of my nitrogen and phosphate waste if you give me some of that fixed carbon"). The mutualism can enjoy a substantial selective advantage over the corresponding non-mutualized host, so the negotiated settlement will rapidly spread, and selection will then occur for an increasingly efficient mutualism—effectively better pay and conditions in exchange for higher worker productivity.

One consequence of a good relationship between "employer" (host) and "employee" is that there are almost certain to be metabolic redundancies on both sides, so, if the mechanisms for achieving metabolic exchange are relatively simple, selection may then favor closer metabolic integration. Why should the host continue to synthesize amino acid X if the symbiont can synthesize an oversupply and can leak some of this to the host? If the leakiness exists or comes about during the "negotiation period," then the host may in the short term downregulate the corresponding biosynthetic enzymes but, in the longer term (i.e., over evolutionary time), lose the enzymes altogether. Thus the relationship becomes an obligate one on the host side. Modern (shallow water) reef-building corals absolutely require the dinoflagellate endosymbiont *Symbiodinium*. Although many aspects of this relationship remain unclear (see Chap. 8), in the case of the genus *Acropora* (staghorn corals; the dominant corals in the Indo-Pacific), all the available evidence points to a metabolic deficiency in the host pathway of cysteine biosynthesis. This may reflect a step toward closer metabolic integration in this important coral genus. Metabolic cooperation occurs widely among microorganisms, generally in response to selection imposed by resource-poor habitats. Besides the few findings reported above, little is known about shared metabolic pathways (i.e., metabolic pathways in which host and symbiont contribute different enzymatic reactions) in symbiotic associations between cnidarians and their eukaryotic and prokaryotic symbionts. Future experiments will show whether metabolic coevolution is another hallmark of early emerging metazoans. In general terms, the incidence of shared metabolic pathways in animal–microbial symbioses is largely unstudied but may be widespread, albeit not universal, particularly in associations involving symbionts with reduced genomes.

An important question is, to what extent does selection operate independently on components within a holobiont? The answer depends on the degree of integration between the component organisms—if the coupling is weak (i.e., two partners collaborate with limited benefit to each), then selection operates in large part on the components individually. If the association is obligate, then selection is at the holobiont level. The situation is more complex in intermediates between these extremes.

Green hydra can survive without its *Chlorella* partner, but the reciprocal is not true; in this case, selection on the alga is at the level of the holobiont, but for the animal, selection does not exclusively operate at this level. In reef-building corals, the situation is reversed (the survival of the coral animal requires the presence of endosymbiotic *Symbiodinium*, but the dinoflagellate can grow independently), relationship between animal and (dinoflagellate) photosynthetic partner is both more diverse and more complex. Many strains of *Symbiodinium* exist, varying enormously in net benefit that they confer to the host, and some corals transmit symbionts via the eggs, whereas others must acquire them during development. Thus the extent to which selection operates at the level of the holobiont may vary enormously even within a single lineage such as corals. In relatively few cases are animal–bacterial mutualisms understood sufficiently well for predictions to be made on the partitioning of selection pressures, but with the advent of novel sequencing and metabolomics techniques, perhaps this will change in the next few years.

5.4 Rethinking the Role of Immunity

As we have seen in this chapter, early emerging metazoans have an elaborate innate immune system. A conserved TLR-signal transduction cascade is used to sense the presence (or absence) of the mostly beneficial microbes. Downstream of this cascade are taxon-specific effector molecules such as antimicrobial peptides (AMPs). AMPs are either constitutively expressed or inducible by a number of different biotic and abiotic environmentally derived factors. While microbial exposure is certainly a major player in the daily life of an early emerging metazoan, pathogens seem not to be abundant. In fact, up to know we have not yet discovered a single bona fide pathogenic bacterium which is threatening *Hydra* polyps in their natural habitat. Of course, there are pathogens such as *Pseudomonas aeruginosa* strains which can be used in infection experiments to kill *Hydra* polyps. But our molecular fishing expeditions so far have not detected any of them in a natural setting where *Hydra* occur. So, why are animals such as *Hydra* in the absence of obvious pathogens investing so much in the maintenance of an elaborate immune system? Our studies support the view that immune systems evolved as much to manage and exploit beneficial microbes as to fend off harmful ones (Bosch 2014). Evidence for this view comes from at least two directions. First, the discovery that individuals from different species differ greatly in their microbiota and that individuals living in the wild are colonized by microbiota similar to that in individuals grown in the lab points to the maintenance of specific microbial communities over long periods of time. And second, the fact that the number of specific recognition elements and effector components of the innate system in *Hydra* is so high provides a mechanism for keeping track of a large number of microbial partners.

As a result of the finding that interactions between animals and microbes are not specialized occurrences but rather are fundamentally important aspects of animal biology and that AMPs and other components of the immune system are key factors for allowing the right microbes to settle and to kick the less desirable ones out, the view of the role of the immune system has changed radically in the last decade. We now consider the innate immune system simply as the hardware for a functioning interspecies network. Invertebrates—which comprise more than 96 % of animal species—rely on this hardware exclusively without the benefit of an adaptive immune system. Invertebrates are a success story in evolution. So, why do vertebrates in addition to the innate immune system have evolved an adaptive immune system? Several years ago, Margaret McFall-Ngai has proposed (Nature 445, 153) that the memory-based adaptive immune system in vertebrates similar to the innate immune system may have evolved because of the need to recognize and manage complex communities of beneficial microbes. At the time when the paper was published in 2007, this was a very provocative idea. Many have thought the answer lies with lifestyle because some invertebrates are small and short-lived and may not need a memory-based immune system that is suited to the long haul. But as Margaret McFall-Ngai has pointed out, numerous invertebrates are large, have only a single offspring each year, and live a long life. *Hydra* certainly is a good example and considered "non-senescent" or even immortal. Clearly, "long-life" lifestyle features can evolve in the absence of an adaptive immune system. Today, mounting evidence supports McFall-Ngai's early vision and shows not only that the adaptive immune system is modulated by the microbiota but also that components of the adaptive immune system such as different populations of T cells are capable to affect the composition of the microbiota very efficiently. Invertebrates as well as vertebrates are mobile ecosystems, carrying a large number of microbes in their epithelia. It seems that they all use their immune systems as a versatile microbial-management strategy.

Conclusion

Work in early emerging metazoans such as *Hydra* and corals has shown that these seemingly simple creatures provide us with important information in understanding the evolution of epithelial-based innate immunity. The work has contributed to a paradigm shift in evolutionary immunology: components of the innate immune system with its host-specific antimicrobial peptides and a rich repertoire of pattern recognition receptors appear to have evolved in early branching metazoans because of the need to control the resident beneficial microbes rather than because of invasive pathogens. Yet in spite of all these insights in an ultimately simple "holobiont," we have still not been able to coherently integrate the accumulated abundance of information into a truly mechanistic understanding of host–microbe interactions. Questions to be addressed in the future include: What are the functional roles of all members of the *Hydra* holobiont? How do bacteria contribute to the phenotypic stability of their hosts? What effect does a changing environment have on microbial associates and the fitness of the holobiont? Elucidating these issues will not only contribute to the understanding of host interactions with microbial communities in one of the simplest possible animal systems but may also provide conceptual insights into the complexity of host–microbe interactions in general. It seems that we have only begun to understand the true impact of microbiota on shaping host physiology and emergence of new functions, including immunity.

References

Augustin R, Fraune S, Bosch TCG (2010) How Hydra senses and destroys microbes. Semin Immunol 22:54–58

Barneah O, Benayahu Y, Weis VM (2006) Comparative proteomics of symbiotic and aposymbiotic juvenile soft corals. Mar Biotechnol (NY) 8(1):11–16

Bosch TCG (2013) Cnidarian-microbe interactions and the origin of innate immunity in metazoans. Ann Rev Microbiol 67:499–518, pdficon

Bosch TCG (2014) Rethinking the role of immunity: lessons from Hydra. Trends Immunol 35(2014):495–502

Bosch TCG, Augustin R, Anton-Erxleben F, Fraune S, Hemmrich G, Zill H, Rosenstiel P, Jacobs G, Schreiber S, Leippe M, Stanisak M, Grotzinger J, Jung S, Podschun R, Bartels J, Harder J, Schroder JM (2009) Uncovering the evolutionary history of innate immunity: the simple metazoan Hydra uses epithelial cells for host defence. Dev Comp Immunol 33:559–569

Lange C, Hemmrich G, Klostermeier U, Miller DJ, Rahn T, Weiss Y, Bosch TCG, Rosenstiel P (2010) Complex repertoire of NOD-like receptors at the base of animal evolution. Mol Biol Evol 28:1687–1702

McFall-Ngai M (2007) Adaptive immunity: care for the community. Nature 445(7124):153

Nyholm SV, Graf J (2012) Knowing your friends: invertebrate innate immunity fosters beneficial bacterial symbioses. Nat Rev Microbiol 10(12):815–827

Voolstra CR, Schwarz JA, Schnetzer J, Sunagawa S, Desalvo MK, Szmant AM, Coffroth MA, Medina M (2009) The host transcriptome remains unaltered during the establishment of coral-algal symbioses. Mol Ecol 18(9):1823–1833

Role of Symbionts in Evolutionary Processes 6

This chapter will look at the different ways in which animals use their microbial symbionts to adapt to the conditions in the environment that they live in. The newfound awareness of a world of complex interactions between developing organisms and the biotic and abiotic components of their environment and of the dependency of phenotypes on other species and environmental conditions presents additional layers of complexity for evolutionary theory and raises many questions that are being addressed by new research programs.

6.1 Microbes as the Forgotten Organ

The human gastrointestinal tract is home to trillions of bacteria comprising thousands of species. In a human individual, this huge bacterial load amounts to 2 kg of bacteria in the gastrointestinal tract. The microbiome is largely defined by two bacterial phylotypes, Bacteroidetes and Firmicutes, with Proteobacteria, Actinobacteria, Fusobacteria, and Verrucomicrobia phyla present in relatively low amounts. Colonization of the infant gut commences at birth when delivery exposes the infant to a complex microbiota. The initial microbiome has a clear maternal signature. Life in the mammalian gut certainly is not easy. However, despite the continuous flow of colonic material and the possible opportunity for microbes and bacteria to get constantly washed out, the bacterial density and diversity in the colon is high and remains relatively and remarkably stable. Existing data indicate that babies born by caesarian section develop a different microbiota with aberrant short-term immune responses and a greater long-term risk of developing immune diseases (Cho and Norman 2013). The microbiome of unweaned infants is simple with high interindividual variability. Next-generation sequencing revealed that the numbers and diversity of strict anaerobes increase as a result of diet and environment, and after 1 year of age, a complex adult-like microbiome is established. Homeostatic mechanisms within the microbiota become less effective in the elderly, and it is clear that the microbiota diverges between those who age healthily and those whose health deteriorates with age. How is the constancy of this complex community structure maintained? And what is it good for? Today the gut represents an important and challenging system for exploring (i) how microbial communities become established within their hosts, (ii) how their members maintain stable ecological niches, and (iii) how these dynamics relate to host health and disease. Due to the documented capacity of the microbiota to produce a diverse range of compounds that play a major role not only in nutrition but also in regulating the activity of distal organs including the brain, the gut microbiota is considered "the neglected endocrine organ" (Clarke et al. 2013, 2014).

6.2 Developmental Symbiosis

Organisms are constructed, in part, by the interactions that occur between the host and its persistent symbiotic bacteria. Although once thought to be exceptions to normal development, such developmental symbioses appear to be ubiquitous among plants and animals. Recent studies document that developmentally active symbionts provide selectable genetic variation for the entire animal and that they may even provide mechanisms for the reproductive isolation that can potentiate speciation.

More than just the product of coevolution, the holobiont is a performance of co-development. In numerous animals, symbiotic interactions are essential to development. For example, bacterial symbionts are essential for the metamorphosis of many invertebrates, for the formation of ovaries by the wasp *Asobara*, and for the germination of orchids. The anterior–posterior axis of the nematode *Brugia malayi* is generated with the help of *Wolbachia* bacteria. If these bacteria are eliminated from the egg, the anterior–posterior polarity fails to develop properly. It is possible that all animals form some of their organs through symbiosis.

As the intestines of germ-free mice (i.e., mice bred in sterile facilities and having no contact with microbes) can initiate, but not complete, their differentiation, bacteria also provide developmental signals to the intestinal epithelia. For complete gut development in mice, the presence of microbial symbionts is required. But mammals aren't the only animals whose gut development depends on microbial symbionts. In zebra fish, microbial symbionts use the beta-catenin signaling pathway to initiate cell division in the intestinal stem cells. In the absence of microbiota, the zebra fish have smaller and less functional intestines, with a paucity of enteroendocrine and goblet cells. All of these defects can be reversed by the introduction of bacteria later in the zebra fish's development. Gene expression profiling of germ-free mice and zebra fish reveals striking parallels in their transcriptional responses to microbiota, with marked changes in the expression of genes involved in cell proliferation, nutrient utilization, and immune function. Such similarities in response to the absence of microbes in fish (zebra fish) and mammalian (mouse) model systems show that animals possess a conserved program of interactions with the symbiotic microbes with which they have coevolved.

Pioneering work led by Margaret McFall-Ngai and Edward Ruby (1991, 2012) has shown that morphogenesis of the light organ of the Hawaiian bobtail squid

Euprymna scolopes is actively induced by *Vibrio fischeri*, one member of the complex seawater microbial community McFall-Ngai (2002). The light organ fails to mature in squids raised without *V. fischeri*, implicating a specific association between *V. fischeri* and the squid. When *V. fischeri* cells are present, they associate with the host along its superficial epithelium. Transcriptomic analyses has revealed that the host recognizes and responds to the presence of these bacteria by expressing a chitinase that primes the bacteria to migrate by chemotaxis up a chitobiose gradient into host tissues. Over the distance of a couple of hundred microns, the squid–*Vibrio* symbiotic system orchestrates a dramatic "winnowing" from thousands of environmental bacterial species interacting with the surface of the light organ to the presence of just one or two strains of *V. fischeri* in the deep organ's crypts. Once in the crypts, *V. fischeri* cells generate the light organ and, through their cell wall peptidoglycans and lipids, induce the loss of the superficial ciliated fields that had facilitated their colonization (McFall-Ngai 2014, 2015).

6.3 The Role of Symbionts in Evolutionary Processes

Developmental symbiosis has been implicated in facilitating major transitions in evolution. Symbiosis may be also responsible not only for the origin of eukaryotic cells but also for the origin of new mammalian cell types and—as discussed earlier in chapter three—for the origin of multicellularity itself.

The endosymbiotic theory of eukaryotic cell formation holds that the origin of eukaryotic life began through the merging of Archaea and bacterial cells and genes. Similarly, animal multicellularity might have emerged from the symbiosis of a choanoflagellate protist with a particular bacterial partner. Nicole King, Howard Hughes Investigator and Professor at the University of California, Berkeley, is tracing the origin of animals by deducing the genetic and developmental foundations of animal origins from shared elements among extant animals and their protozoan relatives, the choanoflagellates (Alegado and King 2014). Choanoflagellates, considered the sister group to animals, can produce unicellular or colonial "morphotypes" in response to certain bacteria (see Chap. 3). In the choanoflagellate *Salpingoeca rosetta*, a sulfonolipid signaling molecule produced by a bacterium from the Bacteroidetes group is sufficient to trigger rosette colony formation. The multicellular aggregates have cytoplasmic connections between their cells as well as a new extracellular matrix around them. They are not loose colonies but appear to be multicellular organisms.

In addition to being obligate inducers of body parts, symbionts play other critical roles to evolution and development. Microbial symbionts form a second type of genetic inheritance, being acquired either through the egg or from the maternal environment. Genetic variation in symbionts can provide phenotypic variation for the holobiont as shown recently in aphids. The pea aphid, *Acyrthosiphon pisum* (Fig. 6.1), is a common agricultural pest that feeds on legume sap, but its survival depends on a symbiotic bacterium, *Buchnera aphidicola*, which synthesizes nutrients unavailable in plant sap. The bacterium has a vastly reduced genome that likely evolved over more than 100 million years of adaptation to the intracellular environment of its host and precludes independent living. Since in aphids symbiotic bacteria provide selectable allelic variation (such as thermotolerance, color, and parasitoid resistance) that enables some holobionts to persist better under different environmental conditions, the close symbiotic association has inspired numerous investigations into how *B. aphidicola* might influence the pea aphid's phenotype and ecology. Is the ability of the holobiont to reproduce in hot weather, having cryptic coloration, or surviving a parasitoid wasp infection, dependent not only on the host's genome but on the genomes of its symbionts? Is the symbiotic bacterium not only necessary for the development of the holobiont, but does its presence or absence determine the holobiont phenotype? To answer these questions and to better understand this symbiotic relationship, Nancy Moran and Yueli Yun from the University of Arizona isolated a *Buchnera* strain that displayed high heat tolerance and injected the strain into pea aphids in which resident, heat-intolerant *B. aphidicola* had been eliminated by heat exposure. The introduced, heat-tolerant *B. aphidicola* strain survived in the pea aphids and was passed down through succeeding generations. In addition, the heat-tolerant bacterium increased heat tolerance in its aphid hosts. Intriguingly, aphids with the *Buchnera* replacement showed a significant increase in their heat tolerance, demonstrating not only the presence of a crosstalk between aphids and their new *Buchnera* cells but also an effect of symbiont genotype on host ecology (Moran and Yun 2015).

Another evolutionary transition, the origin of placental mammals, also may have been permitted and promoted by symbiosis, namely, the incorporation of retroviruses from other organisms. These retroviruses, which contain their own enhancer elements, appear to have allowed the rewiring of cell circuitry to produce the progesterone-responsive uterine decidual cell.

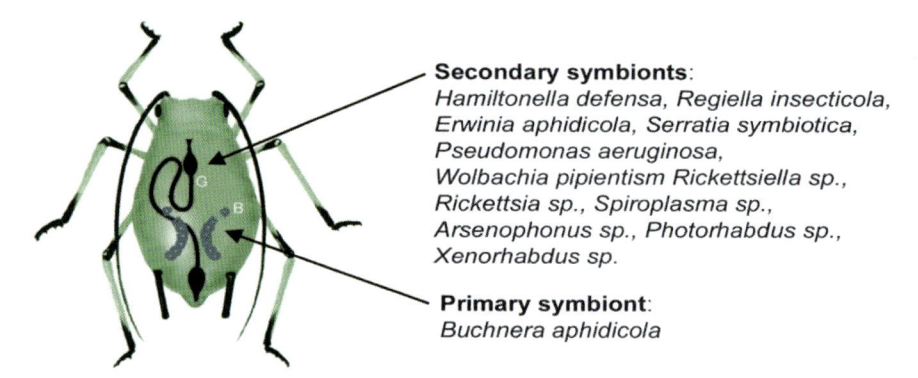

Fig. 6.1 One of the most studied example of symbiotic interaction are aphids and their symbiont *Buchnera aphidicola* that they host in specialized cells called bacteriocytes. Each aphid may host about 6 millions of *Buchnera* cells that are involved in the continuous overproduction of tryptophan and other amino acids (Taken from https://theaphidroom.wordpress.com/tag/buchnera/)

6.4 *Nematostella*, an Early Metazoan Model to Understand Consequences of Host–Microbe Interactions for Rapid Adaptation of a Holobiont to Changing Environmental Conditions

Over the past 20 years, the starlet sea anemone, *Nematostella vectensis*, a small estuarine cnidarian (Fig. 6.2), has emerged as a powerful model system for field and laboratory studies of development, evolution, genomics, molecular biology, and toxicology. *Nematostella* polyps inhabit sediments of brackish water, salt marshes, and lagoons and are commonly found along the east coasts of the North American continent as well as at the British east coast and occupy a large range of environmental conditions concerning temperature and salinity. Spawning is induced by a shift in temperature and exposure to light. The ontogeny of *Nematostella* is characterized by two major developmental transitions (Fig. 6.3). The first transition appears approximately 9–20 days post fertilization (dpf) when the planula larvae metamorphose first into a primary polyp with two mesenteries and four tentacles and later into a juvenile polyp with eight mesenteries and several tentacles. These juvenile polyps need up to 6 more months to reach the second major event, the transition form a juvenile to a sexual mature polyp. These characteristics, together with the exceptional high adaptive potential to varying abiotic factors, turn *Nematostella* into a suitable model organism to understand how environmental factors affect the composition and function of microbiota and the potential consequences of host–microbe interactions for rapid adaptation of a holobiont to changing environmental conditions.

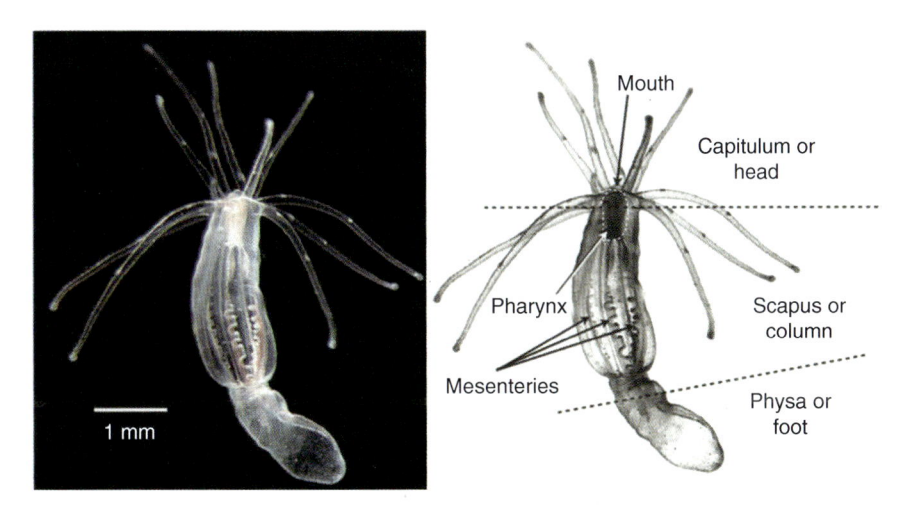

Fig. 6.2 Adult polyp of *Nematostella vectensis*. Field-caught animals are typically ~1 cm long from mouth to tip of the physa, but in culture, well-fed individuals can reach several centimeters in length (Taken from Stefanik et al. (2013))

To investigate the influence of the abiotic factors temperature and salinity on the bacterial composition, a team headed by Sebastian Fraune at the University of Kiel analyzed the diversity of bacterial communities between different environmental conditions at four different developmental time points separately (Mortzfeld et al. 2015). While at planula stage no significant clustering of the bacterial communities could be observed, at later developmental stages, bacterial communities clustered significantly according to environmental conditions. Continuous development under different environmental conditions led to a significant increase in distances between treatments, while the distances within a treatment did not changed significantly over developmental time.

Thus, the composition of the microbiota of each environmental condition becomes more distinct with developmental age.

To identify bacterial groups responding to salinity or temperature shifts in adult polyps, the researchers analyzed the bacterial communities of adult *Nematostella* polyps in more detail. The analyses revealed that samples clustered together based on environmental conditions. Main differences in response to temperature shifts were observed by the increase of Bacteroidetes at 25 °C and the higher abundance of γ-Proteobacteria at 18 °C. Furthermore, six bacterial orders belonging to Bacteroidetes (Flavobacteriales), γ-Proteobacteria (Pseudomonadales, Alteromonadales, and Aeromonadales), β-Proteobacteria (Burkholderiales), and Planctomycetes (Phycisphaerales) responded significantly to changes in salinity in this study. Thus, long-term acclimatization to different environmental conditions is apparently leading to a robust tuning of bacterial colonization. How the regulation of the different bacterial communities is achieved and what the leading

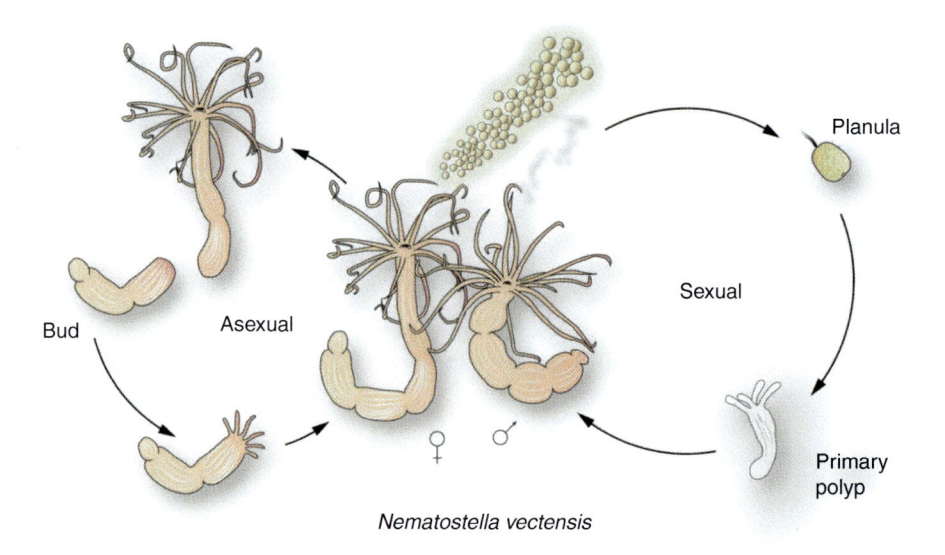

Fig. 6.3 The life cycle of *Nematostella vectensis* (Credit: Iva Kelava and "The Node: the community site for and by developmental biologists http://thenode.biologists.com/)

factors are for the establishment of bacterial communities under different environmental conditions is not known yet. The adjustments in fine-scale community composition following environmental change could represent a way to buffer the impact of environmental changes and thereby contribute to the maintenance of homeostasis. To uncover the potential fitness consequences of this change in bacterial colonization will be a challenging task for the future.

6.5 Rapid Adaptation to Changing Environmental Conditions: The Coral Probiotic Hypothesis

The vast diversity of life on earth is the result of evolutionary processes that acted for billions of years. Consequently, it is often assumed that evolution requires long periods of time. Evolutionary adaptation to new environments as driven by natural selection can, however, occur very rapidly within tens of generations. This raises two questions: First, what are the mechanisms of rapid adaptation? And second, which factors enable and which factors prevent rapid adaptation? The observations described above in *Nematostella* indicate that by changing its microbial community, the holobiont indeed can adapt to changing environmental conditions much more rapidly (weeks to years) than it can via mutation and selection of the host itself (thousands of years). A scheme illustrating how the holobiont can manage changing environmental conditions by integrating an interchangeable set of microbial symbionts is shown in Fig. 6.4.

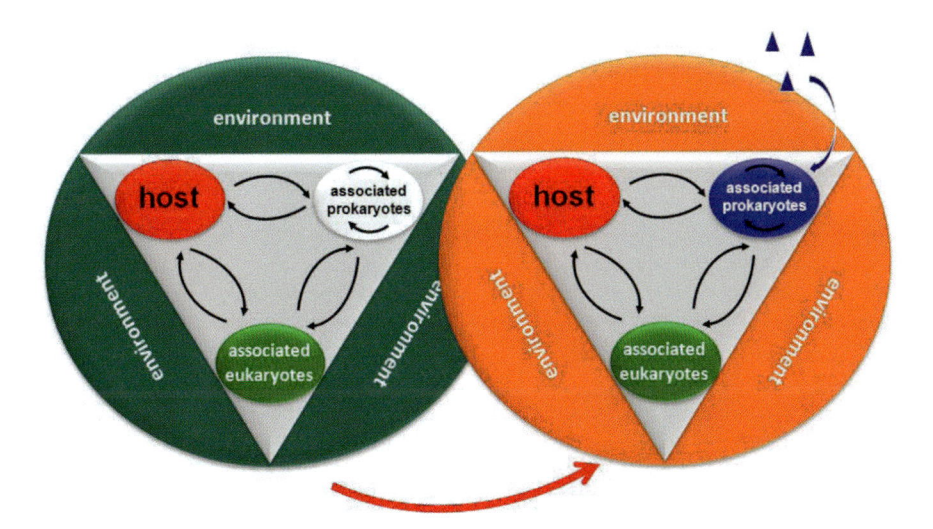

Fig. 6.4 The holobiont in changing environmental conditions. The integration of microbial symbionts allows animals to populate ecological niches that would otherwise be inaccessible to the host. The strict interdependence of a metazoan host with its diverse microbial associates emphasizes the importance of cross-kingdom interactions in species and lineage evolution

The ability to adapt to stresses such as high temperature and infection by specific pathogens rather rapidly has led to the Coral Probiotic Hypothesis Reshef et al. (2006). In a seminal paper published in Environmental Microbiology in 2006, researchers from Tel Aviv University including Leah Reshef, Omry Koren, Yossi Loya, Ilana Zilber-Rosenberg, and Eugene Rosenberg proposed that a dynamic relationship exists between symbiotic microorganisms and corals at different environmental conditions, which bring about a selection for the most advantageous coral holobiont. According to the "hologenome theory" published in 2008 by Ilana Zilber-Rosenberg and Eugene Rosenberg (2008), most animals inherit much the same microbes as the previous generation, and closely related species will have closely related microbiomes. Importantly, changes in the microbiome—from a shift in the ratio of different microbes to the acquisition of new ones—can allow the holobiont to adapt quickly to changing circumstances and even acquire new abilities during its lifetime Rosenberg et al. (2007, 2009). Seth Bordenstein and Robert Brucker from Vanderbilt University in a perspective paper written for the Journal *"ZOOLOGY"* (Fig. 6.5) have nicely illustrated that the hologenomic concept of a species basically is that the beneficial microbiome is an extension of the animal species' genome and mitochondria ("Mt" in Figure 34). If the hologenome theory stands up, it would mean that it's not just physical barriers like mountains and oceans that can separate closely related species and stop them breeding but their microbes as well. This does not mean that the Darwinian model of evolution is wrong. But it does mean that we have to rethink the significance of microbes. As any theory with challenging content, Rosenberg's "hologenome theory" is receiving not only applause but also disgruntled objections. Of course a productive theory not only requires painstaking thought but also supportive evidence. But, as August Weismann once wrote to his friend and colleague Ernst Haeckel on the occasion of receiving a copy of Haeckel's *"Gastrae-Theorie,"* "*it*

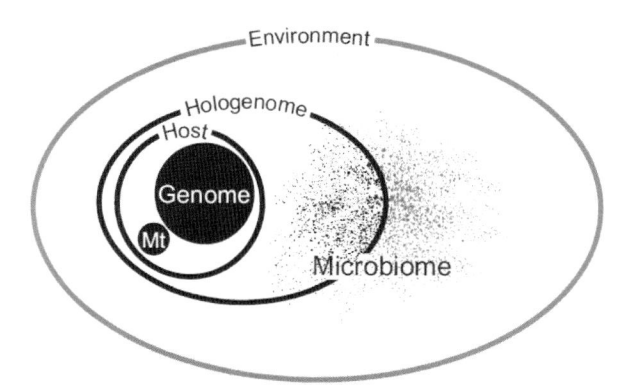

Fig. 6.5 A simplified diagram of the hologenome. The hologenomic concept of a species is that the beneficial microbiome (*black*) is an extension of the animal species' genome and mitochondria (Mt). That beneficial microbiome is deterministically acquired. Although there is some plasticity in the host-associated microbiome derived from the environment (such as diet and abiotic factors, *gray*), the host limits the potential members of the total microbiome (Taken from Brucker and Bordenstein (2013a, b), Zoology)

is so easy to present no theories under the pretense that the necessary foundation of facts is still missing. As though theory must not show what facts are to be looked for. Whether it is correct or not, the facts to be discovered must teach us" (cited from Weismann to Haeckel, January 29 1874).

Seth Bordenstein and Kevin Theis recently have published a well-written essay (PLoS Biology 2015) in which they argue for the terms "holobiont" and "hologenome" to be used to account for the networks that comprise a host and its microbiota. In this essay, the authors clarify what these terms mean, and what they do not, and place them in the historical context of studies of biology and genomics. In the manuscript, the "holobiont/hologenome" concept is explained in 10 principles, illustrated with relevant examples from a broad range of fields of biology. The principles summarize how our understanding of fundamental aspects of biology are being impacted by the changing realization that organisms also include their microbiota and that the microbiome is far more important than previously thought in evolutionary biology and may turn out to be as essential as the nuclear genome. By cooperating with fast-evolving microbes, animals can rapidly adapt to changing environmental conditions. As shown in the next chapter, microbial symbionts may also play a fundamental role in the origin of new species.

6.6 The Role of Symbionts in Speciation

Reproductive isolation is a sine qua non of speciation, and recent evidence suggests that symbiotic microbes may facilitate such isolation Brucker and Bordenstein (2012). Studying the basis of reproductive isolation in three related wasp species, Brucker and Bordenstein (2013a, b) have identified gut microbiota as a cause of hybrid lethality. The wasp species *Nasonia giraulti* and *Nasonia longicornis* have a similar array of gut bacteria and can produce healthy hybrid offspring, but when either wasp mates with the more distantly related *Nasonia vitripennis*, which has different gut microbes, their hybrid offspring die. In contrast, when the hybrid offspring of *N. vitripennis* and each of the other wasp species are raised in a germ-free environment, their hybrid offspring survive. Furthermore, when germ-free offspring of *N. vitripennis* are inoculated with the gut microbes from either of the two other parent species, they die. Thus, a mismatch between the hybrid wasps and their inherited microbiomes appears to be lethal. This suggests a possible evolutionary process whereby populations become increasingly reproductively isolated through the divergence of their microbiomes and may lead to the formation of new species.

Another example of symbiont-induced reproductive isolation is the mating preference exhibited by the fruit fly *Drosophila melanogaster*. *Drosophila* strongly prefer to mate with individuals whose larvae were reared on the same diet as they were. This mating preference was abolished after antibiotic treatment and restored after inoculation of treated flies with microbes from the dietary media, indicating that mate choice is determined by microbes rather than diet. Infection of antibiotic-treated flies with the bacterium *Lactobacillus plantarum* led to the induction of mating preferences, probably through the production of different cuticular hydrocarbons, pheromones that are known to influence mating.

Thus, symbionts are critical to normal development and evolution. They help generate organs, they can produce selectable variant phenotypes, they can create the conditions for reproductive isolation, and they may be the facilitators of evolutionary transitions. Symbiotic relationships are the signature of life on earth, and evolutionary biology has to include developmental symbiosis as a major component. As Brucker and Bordenstein have stated (2014) *"Biology has entered a new era with the capacity to understand that an organism's genetics and fitness are inclusive of its microbiome"* (cited from Brucker and Bordenstein 2014).

The newly discovered, interactive world of holobionts and instructive environments is a nature that is very different from the biomes seen through the lens of the modern synthesis. Animals are not individuals by the traditional anatomical, physiological, immunological, genetic, or developmental accounts. Rather, developmental symbiosis generates holobionts, organisms composed of numerous genetic lineages whose interactions are critical for the development and maintenance of the entire organism. With these changes comes a shift in how we postulate evolution works. Natural selection may function at the level of the holobiont; genes can sometimes be considered followers, not leaders of phenotypic evolution; and developing organisms can modify their environments and then be modified by them. *Documenting, comprehending, and understanding the ramifications of these phenomena are the areas of the newly emerging field of ecological evolutionary developmental biology* (cited from Gilbert et al. 2015).

References

Alegado RA, King N (2014) Bacterial influences on animal origins. Cold Spring Harb Perspect Biol 6(11):a016162

Bordenstein SR, Theis KR (2015) Host biology in light of the microbiome: ten principles of holobionts and hologenomes. PLoS Biol 13(8), e1002226

Brucker RM, Bordenstein SR (2012) Speciation by symbiosis. Trends Ecol Evol 27:443–451

Brucker RM, Bordenstein SR (2013a) The hologenomic basis of speciation: gut bacteria cause hybrid lethality in the genus Nasonia. Science 341:667–669

Brucker RM, Bordenstein SR (2013b) The capacious hologenome. Zoology (Jena) 116(5):260–261

Brucker RM, Bordenstein SR (2014) Response to Comment on "The hologenomic basis of speciation: gut bacteria cause hybrid lethality in the genus Nasonia". Science 345:1011

Cho CE, Norman M (2013) Cesarean section and development of the immune system in the offspring. Am J Obstet Gynecol 208(4):249–254

Clarke G, Grenham S, Scully P, Fitzgerald P, Moloney RD, Shanahan F, Dinan TG, Cryan JF (2013) The microbiome-gut–brain axis during early life regulates the hippocampal serotonergic system in a sex-dependent manner. Mol Psychiatry 18:666–673

Clarke G, Stilling RM, Kennedy PJ, Stanton C, Cryan JF, Dinan TG (2014) Minireview: Gut microbiota: the neglected endocrine organ. Mol Endocrinol 28(8):1221–1238

Gilbert SF, Bosch TCG, Ledón-Rettig C (2015) Eco-Evo-Devo: developmental symbiosis and developmental plasticity as evolutionary agents. Nat Rev Genet 16(10):611–622

McFall-Ngai MJ (2002) Unseen forces: the influence of bacteria on animal development. Dev Biol 242:1–14

McFall-Ngai MJ (2014) The importance of microbes in animal development: lessons from the squid-vibrio symbiosis. Annu Rev Microbiol 68:177–194

References

McFall-Ngai MJ (2015) Giving microbes their due–animal life in a microbially dominant world. J Exp Biol 218(Pt 12):1968–1973

McFall-Ngai MJ, Ruby EG (1991) Symbiont recognition and subsequent morphogenesis as early events in an animal-bacterial mutualism. Science 254(5037):1491–1494

McFall-Ngai MJ, Ruby EG (2012) Deciphering the language of diplomacy: give and take in the study of the squid-vibrio symbiosis. In, Microbes and Evolution: the World that Darwin Never Saw, eds. S Maloy, R Kolter, ASM Press. pp 173-180.

Moran NA, Yun Y (2015) Experimental replacement of an obligate insect symbiont. Proc Natl Acad Sci U S A 112:2093–2096

Mortzfeld B, Urbanski S, Reitzel AM, Künzel S, Technau U, Fraune S (2015) Response of bacterial colonization in Nematostella vectensis to development, environment and biogeography. Environ Microbiol. doi:10.1111/1462-2920.12926

Reshef L, Koren O, Loya Y, Zilber-Rosenburg I, Rosenberg E (2006) The coral probiotic hypothesis. Environ Microbiol 8(12):2068–2073

Rosenberg E, Koren O, Reshef L, Efrony R, Zilber-Rosenburg I (2007) The role of microorganisms in coral health, disease and evolution. Nat Rev Microbiol 5:355–362

Rosenberg E, Sharon G, Zilber-Rosenburg I (2009) Opinion: the hologenome theory of evolution contains Lamarckian aspects within a Darwinian framework. Environ Microbiol 11(12):2959–2962

Stefanik DJ, Friedman L, Finnerty JF (2013) Collecting, rearing, spawning and inducing regeneration of the starlet sea anemone, Nematostella vectensis. Nat Protoc 8:916–923

Weismann to Haeckel, 27 January 1874, in Georg Uschmann anfd Bernhard Hassenstein, "Der Briefwechsel zwischen Ernst Haeckel und August Weismann", in Kleine Festrede aus Anlass der hundertjährigen Wiederklehr der gründung des Zoologischen Institues der Friedrich-Schiller-Universität Jena im Jahre 1865, ed. M. Gersch (Jena: Friedrich-Schiller-Universität, 1965), pp 35-36

Zilber-Rosenberg I, Rosenberg E (2008) Role of microorganisms in the evolution of animals and plants: the hologenome theory of evolution. FEMS Microbiol Rev 32(5):723–735

The *Hydra* Holobiont: A Tale of Several Symbiotic Lineages

7

The personal journey of one of us (TB) toward the realization that beneficial microbes are important began in summer 2000 when Jens Schröder, a dermatologist at Kiel University, on the occasion of the inauguration of the new chair of Zoology asked if *Hydra* were not a good model system to investigate the biochemistry of epithelial defenses. Nobody at that time would have anticipated that this triggered the development of a novel model system in comparative and evolutionary immunology. Up to that moment, *Hydra* was all for examining developmental mechanisms in an evolutionary context and to uncover basic principles of pattern formation and stem cell regulation. To think of immune reactions as equally important features of an animal did not come to our mind. That evolution of a simple multicellular animal such as *Hydra* means both invention of developmental pathways to shape and maintain a given body plan and also to protect this body all life long, and understanding that part of that context requires understanding the biotic and abiotic environment in which *Hydra* evolved turned out to be enlightening and exciting. This chapter will show just how much we know about host–microbe interactions in *Hydra* and what these findings mean in a more general context of holobiont research.

Hydra represents a classical model organism in developmental biology which was introduced by Abraham Trembley as early as 1744 (Fig. 7.1). Because of its simple body plan, having only two epithelial layer (an endodermal and ectodermal epithelium separated by an extracellular matrix termed mesoglea); a single body axis with a head, gastric region, and foot; and a limited number of different cell types, *Hydra* served for many years as model in developmental biology to approach basic mechanisms underlying de novo pattern formation, regeneration, and cell differentiation.

7.1 Rationale for Studying Host–Microbe Interactions in *Hydra*

Novel computational tools and genomic resources have brought a molecular perspective on the *Hydra* holobiont. The genome sequence of *Hydra magnipapillata* revealed an unexpectedly high genetic complexity. In the sequencing process, we accidentally also sequenced the genome of a bacterial species in the *Curvibacter* genus that is stably associated with *Hydra*. The *Hydra magnipapillata* genome is large (1290 M base pairs in size) and contains approximately 20,000 protein-coding genes (Chapman et al. 2010). We found clear evidence for conserved genome structure between *Hydra* and other animals including humans. This contrasts with organisms such as *Drosophila* and the worm *C. elegans* where gene order has been shuffled extensively during evolution. In spite of the fact, however, that *Hydra* belongs to one of the phylogenetically oldest eumetazoan lineages, this organism certainly is not a "living fossil"; its genome contains a rather unique combination of ancestral, novel, and "borrowed" (e.g., via horizontal gene transfer) genes, similar to the genomes of other animals. For analytical purposes, an important technical breakthrough was the development of a transgenic procedure allowing efficient

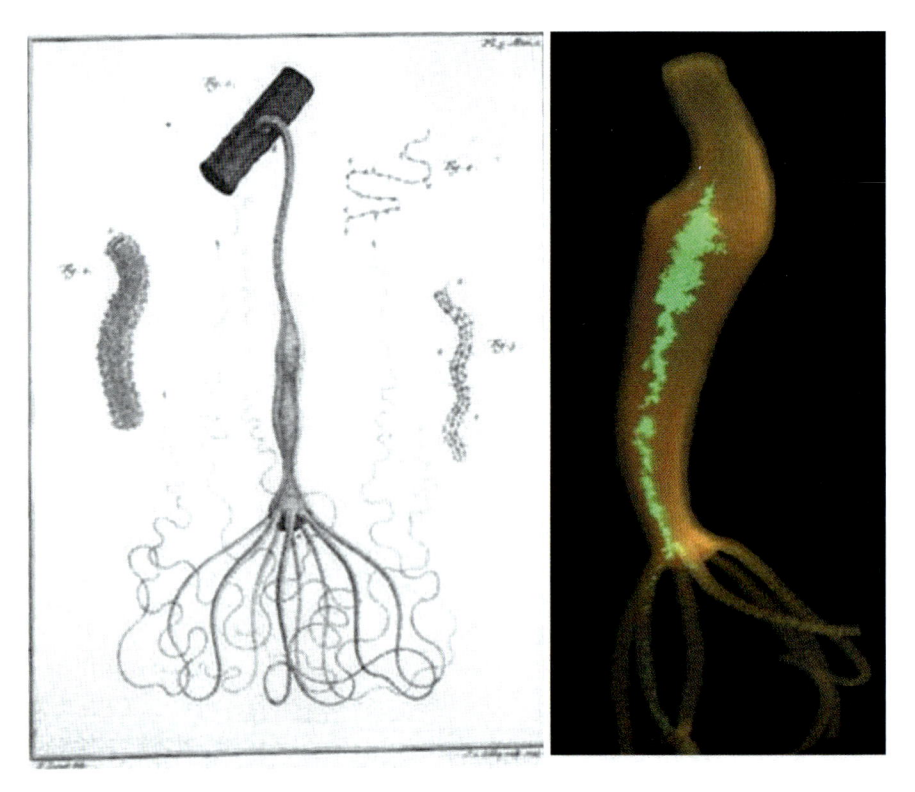

Fig. 7.1 *Hydra* then in 1744 and now. *Left*, taken from Abraham Trembley (1744). *Right*, transgenic polyp containing GFP-expressing stem cells

generation of transgenic *Hydra* lines by embryo microinjection (Fig. 7.1). This not only allows functional analysis of genes controlling development and immune reactions but also in vivo tracing of cell behavior (Wittlieb et al. 2006).

Hints at the richness of life *Hydra* polyps carry in their epithelia came from metagenomic sequencing (Bosch 2013). Next-generation sequencing can provide millions of reads of 100–400 bases at very high speeds and characterize the entire community of bacteria. Sequencing studies are augmented by analysis of the transcriptome and show that individuals from *Hydra* to human are not solitary, homogenous entities but consist of complex communities of many species that likely evolved during a half-billion years of coexistence. Understanding the relationships between the different members of a given holobiont has helped to form an integrated view of an organism. From an evolutionary and ecological perspective, the complexity of these issues at the interface with the microbial environment is invigorating. But within that complexity lie answers to the fascinating question of how trans-kingdom interactions have so profoundly shaped the evolution of life. How to approach this complexity? Given that multiple interconnected factors between both macroscopic host and microbiota (bacteria, archaea, viruses) are at play, it seems prudent to use a genetically tractable and morphologically simple model host. Additionally, an ideal host animal should associate with only a limited number of microbial species, which can be cultured independently, allowing one to deconstruct the complex host–microbe interactions in great detail. Such a model can function as a living test tube and may be a key to dissecting the fundamental principles that underlie all host–microbe interactions. And if such a model represents an ancient animal phylum, it will reveal important insights not only into host–microbe interactions but also into the evolutionary guiding principles underlying the recognition, maintenance, and colonization processes that are also relevant to a mammalian holobiont. One such phylogenetically ancient organism that fits these criteria is *Hydra*.

The *Hydra* holobiont involves at least three types of organisms (*Hydra*, bacteria, and algae) that all share a long coevolutionary history and appear to depend on each other. Interspecies interactions in the freshwater polyp *Hydra* focused on the interaction between symbiotic algae and host cells were already studied decades ago. The research was expected not only to provide insights into the basic "tool kit" necessary to establish symbiotic interactions but also to be of relevance in understanding the resulting evolutionary selection processes. In the meantime the remarkable improvements in DNA sequencing made it evident that in *Hydra* a long-term persistence of mutualistic associations is prevalent not only in two-party interactions of polyp and symbiotic algae but also in more complex systems comprising three or more associates including bacteria and viruses. Thus, studying interspecies interactions in *Hydra* and other cnidarians promises a paradigmatic example of a complex community that influences the host's health and development.

Molecular tools including transgenesis as well as rich genomic and transcriptomic resources and a limited number of microbial symbionts make *Hydra* a valuable holobiont system, which can be manipulated experimentally. Accessibility to both gain-of-function and loss-of-function experiments allows for thorough analysis of molecular pathways. Because of its simplicity in body structure and its exclusive reliance on the epithelial innate immune system, the maintenance of epithelial barriers can be investigated in the absence of the adaptive immune system and other immuno-related cell types and organs. Moreover, due to the relatively simple microbial community structure associating with the animal, which consists of only few bacterial phylotypes (most of which can be cultured in vitro), the influence of the microbiota under healthy and disease condition can be dissected.

Yet, even in a simple animal as *Hydra*, any attempt to understand the complex interplay between the host and its associated microbes is an exceedingly challenging problem due to the complexity of the microbiota and also due to the complexity of the ectodermal and endodermal surface. Overcoming these challenges has required the development of unique in vivo experimental approaches. One of our most important tools in this regard is gnotobiotics ("known life"), a technology involving the use of microbiologically sterile ("germ-free") animals. Germ-free polyps are produced by incubating them for 1 week in an antibiotic solution and culturing them in a sterilized lab facility to keep them free from bacteria. Using germ-free polyps, we can manipulate the *Hydra* holobiont by introducing a single bacterial species ("mono-colonization") or a defined species mixture. We can then combine gnotobiotics with a battery of tools for molecular analysis including transgenics in order to study bacterial–epithelial crosstalk in *Hydra*. Germ-free *Hydra* polyps serve us as platforms for hypothesis testing and finding out the consequences of a good relationship with bacteria. Finally, since nearly all known genes involved in innate immunity are present in *Hydra*, the uncovered basic molecular machinery can be translated to more complex organisms including humans.

7.2 The Hydra Microbiota

In the preceding chapter (Chap. 4), we have seen that bacteria in *Hydra* are specific for any given species and that these animals maintain their specific microbial communities over long periods of time. Closely related *Hydra* species such as *Hydra vulgaris* and *Hydra magnipapillata* are associated with a similar microbial community. In line with this, comparison of the phylogenetic relatedness of *Hydra* species with that of cognate associated bacterial communities reveals a high degree of congruency. Since the composition of the microbiome in *Hydra* parallels the phylogenetic relationships of *Hydra* species, host–microbe interactions in Hydra seem to follow a "phylosymbiotic" pattern (see Chap. 4). But which bacteria are key members of this community?

16S ribosomal RNA (rRNA) sequencing, a common method used to identify and compare bacteria present within a given sample, allowed to uncover the microbial community in a given *Hydra* polyp. Based on this method, the dominant bacteria species (75 % relative abundance) in the laboratory strain *Hydra vulgaris* (AEP) is *Curvibacter* sp. *Curvibacter* is a gram-negative, rod-shaped bacterium from the genus *Curvibacter* and the family of Comamonadaceae which is motile by having polar flagella. In addition to this main colonizer, *Hydra vulgaris* AEP has about 150 operational taxonomic units (OTUs) per polyp according to the chao1 index (based on 97 % similarity), indicating that beside the dominant *Curvibacter* population, bacterial diversity within the remaining 25 % of bacteria is still quite high. Members of the microbiome include *Duganella* sp., a gram-negative β-Proteobacteria from the genus *Duganella* in the Oxalobacteraceae family with an abundance of 11 %, *Undibacterium* sp. with a relative abundance of 2 %, *Acidovorax* sp. with an abundance of 0.7 %, *Pelomonas* sp. (abundance of 0.2 %), and *Pseudomonas* sp. with an abundance of 0.4 %. Overall, in the different *Hydra* species cultured in the laboratory, 36 bacterial phylotypes represent three different bacterial divisions and are dominated by Proteobacteria and Bacteroidetes. In *Hydra vulgaris* strain Basel, Bacteroidetes are represented with two phylotypes (or operational taxonomic units, OTUs) and β-Proteobacteria with 7 phylotypes including *Polynucleobacter* as the most abundant bacterium. In *Hydra vulgaris* strain AEP, Bacteroidetes are represented with five phylotypes and β-Proteobacteria with 9 phylotypes which include the abundant *Curvibacter* bacterium. In *Hydra oligactis*, the majority of phylotypes belongs to the α-Proteobacteria (Rickettsiales).

The fact that *Hydra* living in the wild in their native habitats are colonized by a composition of microbes similar to that of the *Hydra* polyps cultured under controlled laboratory conditions for extended periods of time indicates that the *Hydra* host selectively shapes its bacterial community. Genetic factors of the host appear to outweigh environmental influences in determining microbial surface colonization. Immediate questions therefore are: What are the specific factors that come into play? And how does epithelial homeostasis affect microbial community structure?

One answer lies in the tissue architecture of the host. Any change in the cellular composition of the epithelium immediately causes changes in the microbiome. Experiments on a mutant strain of *Hydra magnipapillata* which has temperature-sensitive interstitial stem cells showed compelling links between the presence or absence of distinct cell types and the bacterial community. In this mutant strain, treatment for a few hours at the restrictive temperature induces quantitative loss of the entire interstitial cell lineage while leaving both the ectodermal and the endodermal epithelial cells undisturbed. As shown in Fig. 7.2, overall morphology and integrity of the epithelium remains unaffected by the temperature treatment despite of disappearance of the entire interstitial cell lineage including interstitial stem cells, nematocytes, neurons, and gland cells. Intriguingly, these changes in the cell composition cause significant changes in the microbial community. Especially two bacterial phylotypes change drastically due to the treatment. The dominant bacterial phylotype in untreated polyps belonging to the β-Proteobacteria is decreased in treated polyps lacking the interstitial cell lineage. In contrast, a bacterial phylotype belonging to the Bacteroidetes increased in abundance in polyps lacking the interstitial cell lineage compared to untreated and control polyps. Thus it appears that there is direct interaction between the presence of certain cell types in the epithelia and the microbiota.

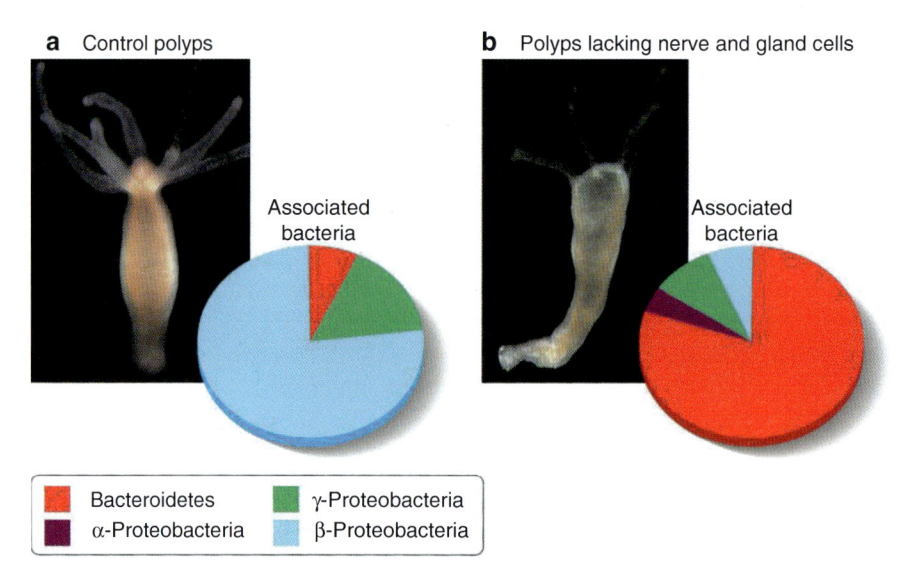

Fig. 7.2 Disturbed tissue homeostasis affects the associated bacterial community. (**a**) Control polyp with its associated bacterial community. (**b**) Polyp lacking nerves and gland cells with its associated bacterial community (Taken from Fraune et al. *Microbe* 2009)

7.3 Linking Tissue Homeostasis, Development, and the Microbiota

Little is known about how developmental pathways and mechanisms controlling tissue homeostasis in the adult are linked to components of the innate immune systems. In an unbiased search for factors maintaining stem cell self-renewal and thereby controlling longevity of *Hydra*, the transcription factor FoxO was found to be strongly expressed in all three stem cell types but silent in terminally differentiated cells. FoxO's well-documented function in life span regulation in other organisms led us to speculate that in *Hydra*, FoxO might be a key driver for the continuous self-renewal capacity of stem cells.

To assess this directly, we performed gain- and loss-of-function experiments. Overexpression of FoxO increased proliferation of stem cells. Silencing of FoxO influenced the delicate balance between stem cells and differentiated cells by increasing numbers of cells going into terminal differentiation, accompanied by a considerable slow down of population growth rate. But then came a surprise. When checking the phenotypes of FoxO knock-down polyps in great detail, Anna Marei Boehm, a graduate student in the Bosch lab, discovered that FoxO downregulation not only resulted in slow growth and increased terminal differentiation of cells. FoxO downregulation also caused drastic changes in the expression level of the AMPs arminin, hydramacin, and periculin2b (Boehm et al. 2012). Consistent with that, in silico analysis revealed multiple FoxO-binding sites on the promoter sequences of the three AMPs. These unanticipated observations indicate that

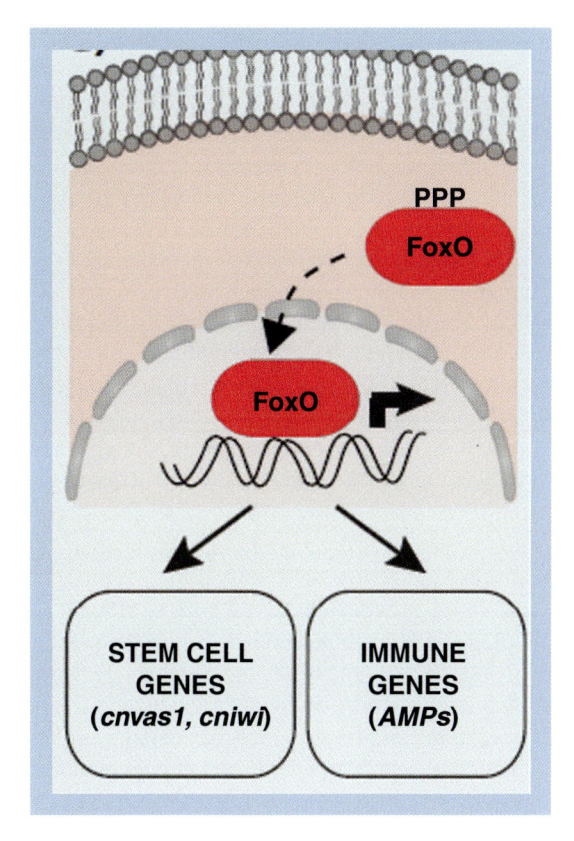

Fig. 7.3 The transcription factor FoxO regulates stem cell and immune genes in *Hydra* (Modified from Boehm et al. 2013)

FoxO-dependent transcriptional programs control not only continuous tissue proliferation but also the synthesis of antimicrobial peptides—and thereby the microbiota composition (Fig. 7.3). Anna Marei Boehm therefore discovered nothing less than a direct link between tissue homeostasis, development, immunity, and the microbiota. This previously unrecognized link is shedding at least some light on the age-old problem of how developmental pathways are linked to components of innate immunity.

The implication of FoxO factors in the regulation of both systems, stem cells and immunity, which are affected by profound age-related changes, seems to be deeply conserved. In mammals, FoxO1 is essential for the maintenance of ESC self-renewal and pluripotency, while FoxO3a is critical regulator of stem cell homeostasis in neural, leukemic, and hematopoietic stem cells (HSCs). At the same time, there are strong data implicating FoxO in maintaining immune homeostasis in mammals and influencing the innate immune system by regulating the activity of antimicrobial peptides. In *Drosophila* flies which are overexpressing dFoxO, direct binding of dFoxO to the regulatory region of the gene encoding the antimicrobial peptide

Drosomycin 2 leads to an induction of AMPs synthesis (Becker et al. 2010). Moreover, studies focusing on centenarians (age 100+ years) also showed that there is a potential relationship between FoxO activity and the immune status. Being distinguished by their exceptional high age which is correlated with a particular sequence version of the *foxO3a* gene, centenarians at the same time show a remarkable good health status and a reduced pro-inflammatory profile as it is normally typical for elderly people.

Hydra polyps are potentially immortal and never experience the gradual changes seen in an aging organism leading toward increased weakness, disease, and death. Surprisingly, *Hydra*'s "eternal developmental youth" appears unconnected to the prevailing environmental conditions. This is certainly true for polyps kept under controlled feeding conditions in the laboratory where it makes no difference whether the animals are fed ad libitum or strictly limited. *Hydra*'s stem cells in individual polyps never lose their capacity for unlimited self-renewal. Although few ecological field studies have been undertaken so far, there is also no evidence that environmental conditions affect number or functionality of *Hydra*'s stem cells in the natural habitat which is characterized by seasonal fluctuations in nutrient abundance. So, why have environmental conditions apparently only limited impact on *Hydra*? How is stem cell proliferation and tissue growth coordinated with nutritional conditions? Do the associated bacteria play a role in protecting the host from seasonal fluctuations in nutrient abundance? We do not know yet the answers. However, by combining genetic manipulation of key components of environment-sensing pathways and controlled and germ-free culture conditions, *Hydra* may emerge as a truly suitable model to study how the interaction between genome and environmental factors affects tissue homeostasis and aging. This research is expected not only enhance our understanding of genetic and environmental influences on the evolutionary conserved processes controlling aging but may also have implications for physicians and scientists concerned with human age-associated diseases (Nebel and Bosch 2012).

7.4 Hydra's Mucus Layer Plays a Key Role in Maintaining the Necessary Spatial Host–Microbial Segregation

A characteristic feature of most animal epithelial cells is a dense carbohydrate-rich layer protruding up to 500 nm from the apical cell surface, referred to as the glycocalyx. The dense layer of transmembrane glycoproteins, glycolipids, and proteoglycans excludes large molecules and organisms to have direct access to the cell surface by steric hindrance, whereas smaller molecules might pass through. Visualization of the glycocalyx in cnidarians is not a trivial task. Since conventional chemical fixation procedures fail to preserve this extracellular structure, it has been ignored and overlooked for a long time. Pilot electron microscopic (EM) analysis of *Hydra*'s ectodermal epithelium revealed a smooth uninterrupted layer covering the whole body with exception of the lowest part of the foot (Bosch et al. 2009). Using high-pressure freezing/freeze substitution (HPF/SF), a team of researchers around Thomas Holstein in Heidelberg and Bert Hobmayer at Innsbruck University resolved the ultrastructure of *Hydra*'s glycocalyx, which appears to consist of five distinct layers. Although the

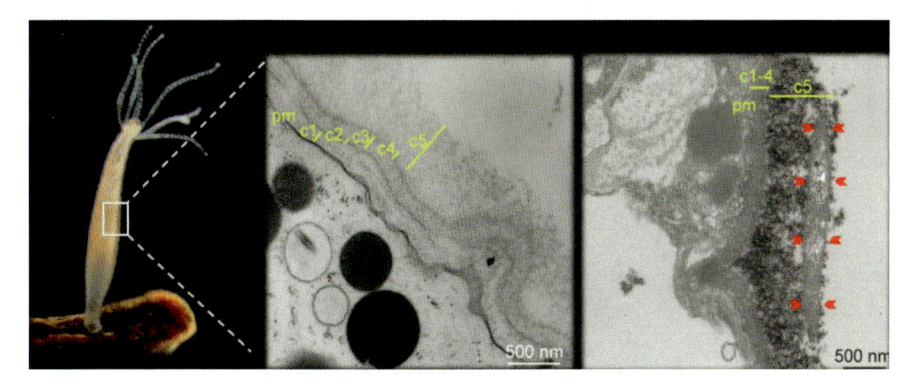

Fig. 7.4 Bacteria colonize the outer layer of *Hydra*'s glycocalyx (Taken from Fraune et al. *ISME J* 2014)

overall structure was termed glycocalyx, at least the outer layer appears to be not membrane bound but resembling a mucus-like gel. This mucus layer is important because it turns out to be the habitat for *Hydra*'s microbial partners. *Hydra*'s stably associated bacteria are almost exclusively associated with the ectodermal epithelium and in particular with the glycocalyx layer. Interestingly, only the outer mucus layer of *Hydra*'s glycocalyx is colonized by bacteria, whereas the dense inner layers seem to remain sterile. This layer, therefore, appears to providing a barrier to both the commensal microorganisms and potential pathogens (Fig. 7.4).

This principle of separation into a habitat for symbiotic bacteria and a physical barrier preventing excessive immune activation was previously described for the mucosal surface of the mammalian colon. In the colon, where most microbes exist, mucus can be clearly subdivided into a tight inner layer composed of tightly stacked mucin polymers and a looser outer layer, as the structure is broken down by proteolysis. Very similar to *Hydra*, the outer mucus layer is colonized with microbes, whereas the inner layer, where the concentration of antimicrobial compounds is high, is relatively sterile and impenetrable—colloquially dubbed the "demilitarization zone." This physical separation, therefore, apparently is a conserved feature which can be traced back to the ancestral metazoan *Hydra*.

Since an intact glycocalyx was found to be required to prevent infection with, e.g., spores from the pathogenic freshwater mould *Saprolegnia ferax* (often called "cotton mould"), *Hydra*'s glycocalyx appears to represents the frontline defense barrier between the external environment and the epithelial tissues.

7.5 Microbes Differ in Embryos and Adult: Embryo Protection

Mammalian embryos are embedded in the uterus, which provides protection during embryogenesis. Many other vertebrates and invertebrates release their oocytes in an environment full of microbes to develop there as "orphan" embryos. How these seemingly unprotected embryos respond to the environment-specific microbial challenge is an interesting albeit not yet understood problem. The most critical phase in the development of any embryo appears to be the period prior to maternal–zygotic transition (MZT) when the embryo starts to utilize its own transcriptional machinery. In this period the cells do not transcribe their own genes as they have only a biphasic cell cycle consisting of only two steps: the mitosis (M) and the synthesis (S) phase. Only after the MZT, when the G1 and G2 phases are added to the cell cycle, the embryo is able to response actively to environmental signals, for example, with production of heat shock proteins. How is bacterial colonization of the early embryo controlled before MZT? For a long time now, there has been a growing awareness of vertebrate developmental biologists for the significance of the so-called "fertilization envelope" in providing microbial protection in early developmental stages. This protection has to be achieved by maternal mechanisms as the early fish embryo is not using its own transcriptional machinery before MZT. While in most invertebrates the nature of the molecules involved in maternal protection are not known yet, the freshwater polyp *Hydra* uses maternally encoded antimicrobial peptides of the periculin family to protect its embryos.

In female *Hydra* (Figure), oocytes differentiate from clusters of interstitial stem cells committed to the female germline and develop into a mature egg which is attached outside the mother (Fig. 7.5). Upon commitment to female gametogenesis, female interstitial cells, often referred to as nurse cells, produce AMPs of the peptide family periculin and store it in vesicles. Within each cluster of interstitial cells, one of the cells develops into an oocyte, while the other nurse cells are phagocytosed and become incorporated into the cytoplasm of the developing oocyte. Condensed nurse cells constitute the bulk of the ooplasm, persist throughout embryogenesis, and provide active AMPs including members of the periculin family for the developing oocyte. Following fertilization, periculin-containing vesicles get released to the surface of the developing embryo (Fig. 7.6).

During embryogenesis the number of bacterial colonizers is increasing in number and changing in composition. For example, the bacterial phylotypes, belonging to the *Pelomonas* group, and those representing *Curvibacter sp.*, are present only in late developmental stages while they appear to be absent in the early embryo (Fig. 7.7). Thus, early developmental stages have a microbiota which is clearly distinct from later developmental stages. Interestingly, the differential colonization is reflected in differences in antimicrobial activity in embryos compared to adult polyps. Beginning with the gastrula stage, i.e., after MZT, *Hydra* embryos express a set of periculin peptides (periculin 2a and 2b) which replaces the maternal-produced periculin peptides 1a and 1b (Fig. 7.7). This shift in the expression within the periculin peptide family represents a shift from maternal to zygotic protection of the embryo. In adult *Hydra* polyps, additional AMPs, including hydramacin and arminin, contribute to the host-derived control of bacterial colonization.

Why *Hydra* embryos appear to select developmental stage-specific microbes is not yet understood. Are the associated beneficial microbes involved in embryo protection? Observations in a number of aquatic animals indicate a protective function of associated symbionts for the early embryo. Preliminary observations in *Hydra* also point to a role that associated bacteria in microbial defense since bacteria-free

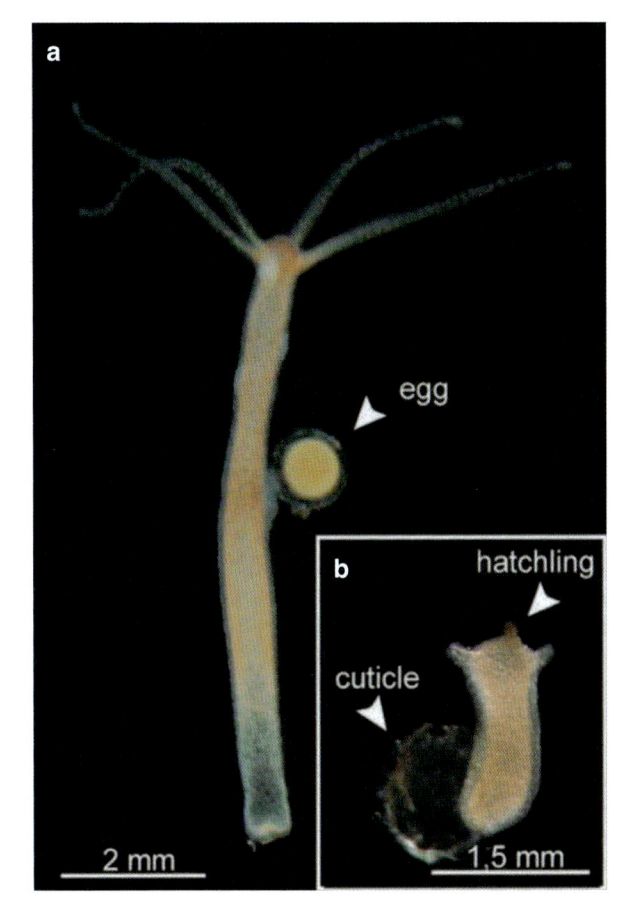

7.5 Microbes Differ in Embryos and Adult: Embryo Protection

Fig. 7.5 (**a**) Female polyp of *Hydra vulgaris* with a developing embryo. *Arrow head* points to developing egg. (**b**) *Hydra* hatchling eclosing from the cuticle. *Arrow heads* point to cuticle and hatchling respectively

polyps and embryos are prone to severe fungal infection, while control animals show no evidence of fungal growth.

How does an animal assemble the specific set of microbes it needs to survive and avoid the microbes that might harm it? We addressed this question by profiling the assembly of the microbiota on *Hydra* epithelium up to 15 weeks post-hatching (Franzenburg et al. 2013). Interestingly, distinct and reproducible stages of colonization can be observed. High initial variability and the presence of numerous different bacterial species are followed by the transient preponderance of the bacterial species that later dominate the adult microbiota. At the end of the colonization process, there is a drastic decrease of diversity. The study which included mathematicians to model the experimental observations (Fig. 7.8) came to the conclusion that both local environmental or host-derived factor(s) modulating the colonization rate and frequency-dependent interactions of individual bacterial community members are important aspects determining the composition of a given microbiome.

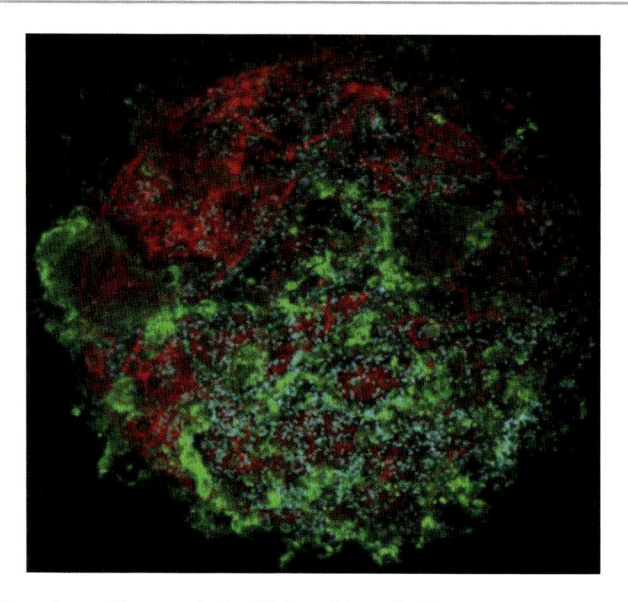

Fig. 7.6 *Hydra* embryo. *Green*, antimicrobial peptide periculin. *Blue*: bacteria. *Red*: actin (Taken from Fraune et al. 2011a)

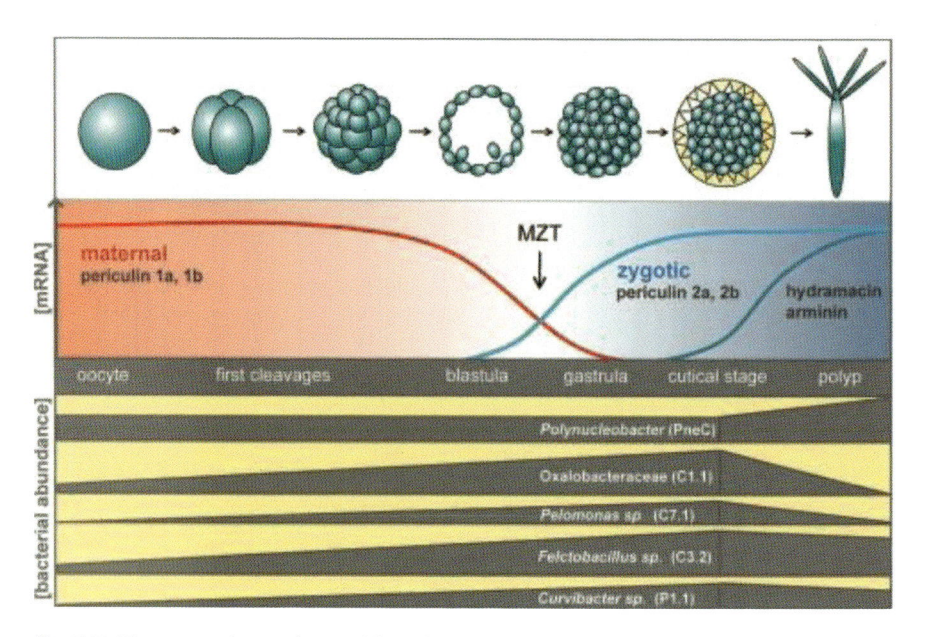

Fig. 7.7 The maternal–zygotic transition of maternally and zygotically produced antimicrobial peptides in the basal metazoan *Hydra* and the corresponding bacterial colonization of the different developmental stages (Taken from Fraune et al. 2011b)

Interestingly, similar trends were observed in a study of the human infant intestinal microbiota. There, profiling the postnatal colonization of 14 human infants by fecal analyses showed that the intestinal microbiota is variable in infants and converges to an adult-like profile with time. Moreover, progression toward the

7.6 Antimicrobial Peptides Function as Host-Derived Regulators of Microbial Colonization 91

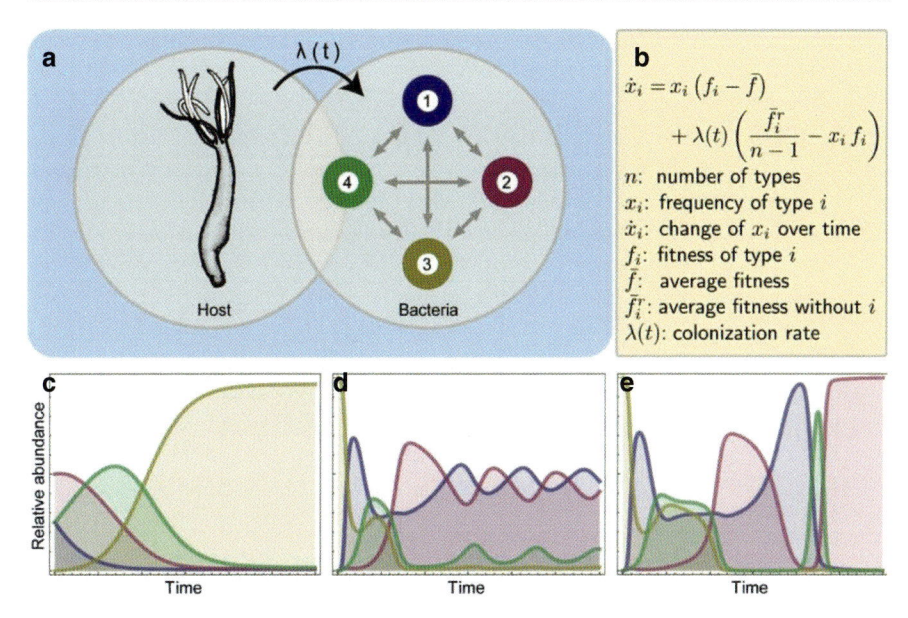

$\dot{x}_i = x_i (f_i - \bar{f})$
$+ \lambda(t) \left(\dfrac{\bar{f}_i^r}{n-1} - x_i f_i \right)$

n: number of types
x_i: frequency of type i
$\dot{x}_i$: change of x_i over time
f_i: fitness of type i
$\bar{f}$: average fitness
$\bar{f}_i^r$: average fitness without i
$\lambda(t)$: colonization rate

Fig. 7.8 Mathematical modeling of the bacterial colonization process in *Hydra* revealed that frequency-dependent growth rates of the bacteria are not enough to explain the dynamics. It needs in addition an external modulation of the colonization process by an environmental (host-derived) factor λ (Taken from Franzenburg et al 2013). (**a**) Mathematical model of the colonization process. (**b**) The basic dynamic equation is the replicator–colonizer equation. (**c–e**) To gain qualitative understanding of the key factors of the colonization process, we followed three successive steps. First, (**c**) bacterial interactions are constant, and type-specific, the colonization rate is constant (**d**) Second, frequency-dependent interactions yield more complex behavior, for example, cyclic patterns under constant colonization rate. (**d**) Third, adding a time-decaying colonization rate, the initial oscillatory/fluctuating behavior then is damped by a factor external to the bacterial community such that the final distribution with a unique single predominant bacterial type is assumed. Thus, frequency-dependent growth rates of the bacteria are not enough to explain the dynamics, an external modulation of the colonization process is additionally required

adult-like microbiota contains a transient state around day 5 which remarkably resembles the stable communities found in older children. Thus, in both *Hydra* and human the progressive development of the adult-like microbial profile seems to require a transient occurrence of the generic adult-like profile.

7.6 Antimicrobial Peptides Function as Host-Derived Regulators of Microbial Colonization

In preceding chapters, we have explained that *Hydra*'s innate immune system and in particular the rich repertoire of antimicrobial peptides produced by ectodermal and endodermal epithelial cells are the major host factors involved in controlling the microbial community. Conventionally, antimicrobial peptides are considered as an effective innate immune weapon against microbial intruders. This view, however, may be a bit simplistic since, for example, ectopic overexpression of antimicrobial periculin1a peptide in *Hydra*'s ectodermal epithelial cells does not cause simple disappearance of the associated microbes but results in a distinct change in the composition of the microbiota (Figure).

Testing the hypothesis of antimicrobial peptides being capable of causing changes in the microbial composition in *Hydra*, we used a transgenic approach and observed that polyps expressing antimicrobial peptide periculin show significant changes in microbiota composition but not in total bacterial numbers (Fig. 7.9). Thus, antimicrobial peptides play a homeostatic role in regulating the makeup of the commensal microbiota in adult *Hydra* polyps. This is consistent with the observations made in the *Hydra* embryos, where maternal AMPs maintain the homeostasis between the bacterial colonizers and the epithelium of the early development stages.

In an extension of this study, we recently addressed the question, if species-specific AMPs sculpture species-specific bacterial communities by selecting for coevolved bacterial partners. Diet, which has a strong influence on the microbiota, and temperature and environment were standardized in our experimental setup. In particular, we investigated the effect of arminin deficiency in *Hydra*. Arminin-deficient and control polyps were inoculated with native as well as foreign bacterial communities characteristic for the closely related species *Hydra oligactis* and *Hydra viridissima*. Whereas control polyps were selected for bacterial communities resembling their native microbiota, this host-driven selection was significantly less pronounced in arminin-deficient polyps. These data provide strong evidence for a role of species-specific AMPs in selecting suitable bacterial partners, leading to host species-specific bacterial associations. These and other observations make it very clear that AMPs function as host-derived regulators of microbial colonization rather than as simple killers.

When considering the function of antimicrobial peptides from an evolutionary perspective, it may be relevant to consider that most if not all AMPs are restricted to a specific genus or even a species and are representing so-called taxonomically restricted genes (TRGs). The presence and use of species-specific effector molecules most likely reflects habitat-specific adaptations to control habitat-specific microbial colonizers. Evolutionary changes in the AMP repertoire of host species would therefore lead to changes in the composition of the associated bacterial community. Since the genetic information encoded by microorganisms can change

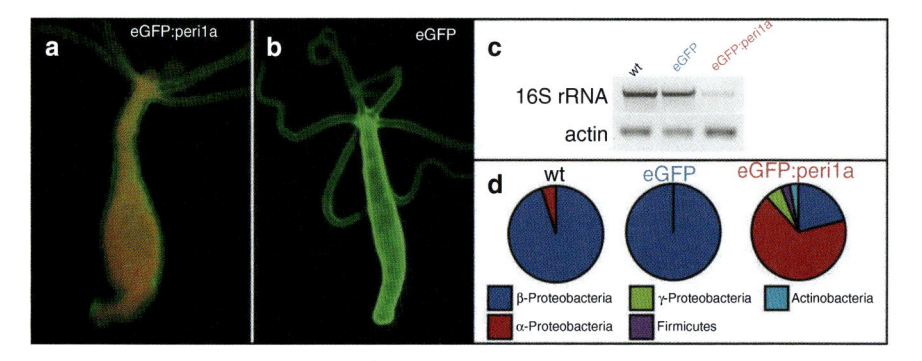

Fig. 7.9 Impact of antimicrobial peptide periculin on microbial composition of *Hydra*. (**a**) Overexpression of eGFP (control) and periculin (**b**) in adult polyps. (**c** and **d**) Change in bacterial colonization after overexpression of periculin (Taken from Fraune et al 2011a)

under environmental demands more rapidly than the genetic information encoded by the host organism, Rosenberg and colleagues suggested in their hologenome theory (see Chap. 6) that changed microbial partners confer greater adaptive potential to environmental changes than alteration and selection processes required for host genome evolution alone.

In sum, there is now convincing evidence that early emerging animal hosts at different developmental stages select for specific microbes by using distinct sets of antimicrobial peptides. These beneficial stage-specific bacteria can have essential roles in antibacterial and antifungal defenses. However, despite the obvious and proven importance of interactions between microbes and their hosts and the fact that hosts control microbial community composition by antimicrobial peptides, little is known at present about the rules that govern the host–microbe assemblies. What are the contributions of species interactions? Is there selection at the community level, and if so, how? Equally pressing questions concern the stability and robustness of within-host microbial communities. A key point is to understand how far the overall function of the microbiota is influenced by individual as well as synergistic contributions of community members. An in-depth understanding of these points requires systematic phenotypic screens in combination with an analysis of the underlying molecular interactions, taking into account the relevant environmental variables.

7.7 Symbiotic Interactions Between *Hydra* and the Unicellular Algae *Chlorella*

In addition to host–bacteria interactions, one species of *Hydra*, *Hydra viridissima*, forms a stable symbiosis with intracellular algae of the *Chlorella* group (Figs. 7.10. and 7.11). Interspecies interactions between symbiotic algae and *Hydra* epithelial cells had

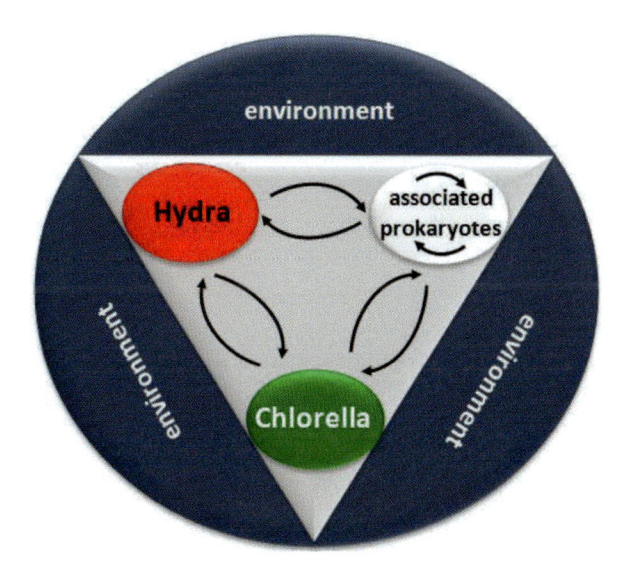

Fig. 7.10 The *Hydra viridissima* holobiont includes not only bacteria but also *Chlorella* algae

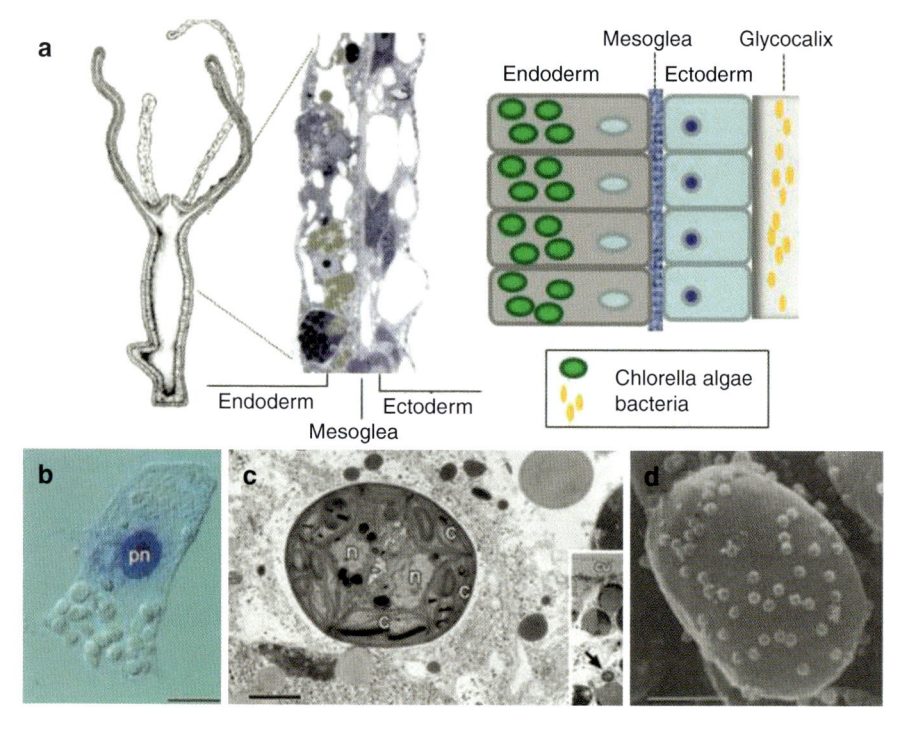

Fig. 7.11 The *Hydra viridissima* holobiont. (**a**) Localization of symbiotic algae and bacteria in *Hydra* epithelium. (**b**) Phase contrast micrograph of a single macerated endodermal epithelial cell containing symbiotic algae in the basal part below the nucleus (*stained blue*). Scale bar, 5 μm (From Habetha et al. 2003). (**c**) Electron micrograph of a symbiotic *Chlorella* within embryonic tissue of *Hydra viridissima* strain A99. The endosymbiont is in the process of mitosis so that two nuclei and several plastids are visible in this section. The inset shows the localization of the alga (*arrow*) within the embryo. Scale bar, 1 μm. *cu* cuticle, *n* algal nucleus, *c* chloroplast. (**d**) *Chlorella* with phytoviruses (Modified from Bosch 2012)

been the subject of research for decades since they not only provide insights into the basic "tool kit" necessary to establish symbiotic interactions but are also of relevance in understanding the resulting evolutionary selection processes (Habetha and Bosch 2005; Bosch 2012). Pioneering studies in the 1980s showed that there is a great deal of adaptation and specificity in this symbiotic relationship (Rahat 1985; Rahat and Reich 1983, 1984; Thorington and Margulis 1981). Symbiotic *Chlorella* are unable to grow independently of the host, indicating a loss of autonomous free-living lifestyle during the evolution of this species. Symbiotic algae also have a substantial impact on sexual reproduction in *Hydra viridissima* by promoting oogenesis. Female gonads almost exclusively are produced only when symbiotic algae are present. Since during oogenesis symbionts are actively transferred from endodermal epithelial cells to the ectodermal oocytes, this oogenesis-promoting role could indicate that the symbionts are critically involved in the control of sexual differentiation in *Hydra viridissima*. Very little is known about the underlying genetics and molecular basis that enables *Chlorella* to survive and proliferate within *Hydra viridissima*'s vacuoles.

7.7 Symbiotic Interactions Between *Hydra* and the Unicellular Algae *Chlorella*

In green *Hydra*, the *Chlorella* symbionts are located in endodermal epithelial cells (Fig. 7.11c). A single endodermal epithelial cell contains usually 20–40 algae (Fig. 7.11d). Each alga is enclosed by an individual vacuolar membrane resembling a plastid of eukaryotic origin like the complex plastids of chlorarachniophytes at an evolutionary early stage of symbiogenesis. Proliferation of symbiont and host is tightly correlated. Although it is not yet known how *Hydra* controls cell division in symbiotic *Chlorella*, the pioneering studies by Rahat and Reich in the 1980s showed that there is a great deal of adaptation and specificity in this symbiotic relationship. Symbiotic *Chlorella* strain A99 is unable to grow outside its polyp host, indicating loss of autonomy during establishment of the intimate symbiotic interactions with *Hydra*. The photosynthetic symbionts provide nutrients to the polyps in the form of maltose or glucose-6-phosphate, enabling *Hydra viridissima* to survive periods of starvation. Symbiotic algae can be removed from *Hydra viridissima* polyps experimentally by various means. During sexual reproduction of the host, *Chlorella* algae are translocated into the oocyte, giving rise to a new symbiont population in the hatching embryo to ensure transmission of the symbiotic algae from generation to generation. Symbiotic algae have severe impact on sexual reproduction in *Hydra viridissima* by promoting oogenesis but not spermatogenesis. This is similar to findings in symbiotic anthozoans where both in a scleractinian coral and in a soft coral, loss of symbionts is correlated with a drastically reduced reproductive output. Very little is known about the underlying genetics and molecular basis which enables *Chlorella* to survive and proliferate within *Hydra viridissima* vacuoles and controls the interaction between both partners. Thus, the *Hydra viridissima* holobiont provides compelling evidence for a complex crosstalk of an ancient epithelial barrier and the residing symbiotic algae and resident bacteria. We currently are exploring the impressive capabilities of this holobiont by asking a range of questions. By what mechanisms does the *Hydra* host recognize its specific algal partner? What are the influences of symbiotic algae on developmental processes of the *Hydra* host and how is the symbiont population maintained in balance over the host's lifetime, such that neither the symbiont overgrow the host nor the host eliminate the symbiont? Do the difficulties of growing symbiotic *Chlorella* outside their host cells reflect the fact that the endosymbionts have transferred some of their genetic material to the nuclear genome of *Hydra*?

Conclusion

Advances in sequencing technologies coupled with new bioinformatic developments have allowed us to start to examine the complex interaction between commensal communities of bacteria residing on the surfaces of *Hydra*'s epithelial barriers. We have uncovered some of the functions of the resident microbiome and have discovered their involvement in innate defenses and maybe even normal development. Despite these insights, however, mechanisms which mediate the interdependent and complex interactions within this holobiont are almost entirely unknown. How do close associations of organisms influence each other's fitness? How do the associated organisms coordinate their interactions at the molecular level? How do the underlying reactive genomes coevolve? The microbiota seems to function as an extra organ, to which the host has outsourced numerous crucial metabolic, nutritional, and protective functions. Yet, what is driving this symbiotic relationship? Do the microbes shape their environment or adapt to the host environment? Clearly, we are far from understanding the *Hydra* holobiont and the numerous interactions between *Hydra*, symbiotic algae, and stably associated bacteria and viruses. But the potential of large-scale gene expression analysis and comparative genome assessments give good reason for hope that elucidating the molecular basis of one of the most remarkable animal–algae–microbe associations formed in millions of years of coevolution is possible in the near future.

References

Becker T, Loch G, Beyer M, Zinke I, Aschenbrenner AC, Carrera P, Inhester T, Schultze JL, Hoch M (2010) FOXO-dependent regulation of innate immune homeostasis. Nature 463(7279):369–373

Boehm AM, Rosenstiel P, Bosch TCG (2013) Stem cells and aging from a quasi-immortal point of view. Bioessays 35(11):994–1003

Boehm AM, Hemmrich G, Khalturin K, Puchert M, Anton-Erxleben F, Wittlieb J, Klostermeier UC, Rosenstiel P, Oberg H-H, Bosch TCG (2012) FoxO is a critical regulator of stem cell maintenance and immortality in Hydra. Proc Natl Acad Sci U S A 109(48):19697–19702

Bosch TCG (2012) What Hydra has to say about the role and origin of symbiotic interactions. Biol Bull 223:78–84

Bosch TCG (2013) Cnidarian-microbe interactions and the origin of innate immunity in metazoans. Ann Rev Microbiol 67:499–518

Bosch, TCG, Augustin R, Anton-Erxleben F, Fraune S, Hemmrich G, Zill H, Rosenstiel P, Jacobs G, Schreiber S, Leippe M, Stanisak M, Grotzinger, Jung S, Podschun R, Bartels J, Harder J, Schroder JM (2009) Uncovering the evolutionary history of innate immunity: the simple metazoan Hydra uses epithelial cells for host defence. Developmental and comparative immunology 33:559–569

Chapman JA, Kirkness EF, Simakov O, Hampson SE, Mitros T, Weinmaier T, Rattei T, Balasubramanian PG, Borman J, Busam D, Disbennett K, Pfannkoch C, Sumin N, Sutton GG, Viswanathan LD, Walenz B, Goodstein DM, Hellsten U, Kawashima T, Prochnik SE, Putnam NH, Shu S, Blumberg B, Dana CE, Gee L, Kibler DF, Law L, Lindgens D, Martinez DE, Peng J, Wigge PA, Bertulat B, Guder C, Nakamura Y, Ozbek S, Watanabe H, Khalturin K, Hemmrich G, Franke A, Augustin R, Fraune S, Hayakawa E, Hayakawa S, Hirose M, Hwang JS, Ikeo K, Nishimiya-Fujisawa C, Ogura A, Takahashi T, Steinmetz PR, Zhang X, Aufschnaiter R, Eder MK, Gorny AK, Salvenmoser W, Heimberg AM, Wheeler BM, Peterson KJ, Böttger A, Tischler P, Wolf A, Gojobori T, Remington KA, Strausberg RL, Venter JC, Technau U, Hobmayer B, Bosch TC, Holstein TW, Fujisawa T, Bode HR, David CN, Rokhsar DS, Steele RE et al (2010) The dynamic genome of Hydra. Nature 464(7288):592–596

Franzenburg S, Fraune S, Altrock PM, Künzel S, Baines JF, Traulsen A, Bosch TCG (2013) Bacterial colonization of Hydra hatchlings follows a robust temporal pattern. ISME J 7(4):781–790

Fraune S, Augustin R, Bosch TCG (2009) Exploring host-microbe interactions in hydra. Microbe 4(10):457–462

Fraune S, Franzenburg S, Augustin R, Bosch TCG (2011a) Das Prinzip Metaorganismus. BIOspektrum 17(6):634–636

Fraune S, Augustin R, Bosch TCG (2011b) Embryo protection in contemporary immunology: why bacteria matter. Commun Integr Biol 4:369–372

References

Fraune S, Anton-Erxleben F, Augustin R, Franzenburg S, Knop M, Schröder K, Willoweit-Ohl D, Bosch TCG (2014) Bacteria-bacteria interactions within the microbiota of the ancestral metazoan Hydra contribute to fungal resistance. ISME J 9(7):1543–1556. doi:10.1038/ismej.2014.239

Habetha M, Anton-Erxleben F, Neumann K, Bosch TCG (2003) The Hydra viridis/Chlorella symbiosis. (I) Growth and sexual differentiation in polyps without symbionts. Zoology 106(2):101–108

Habetha M, Bosch TCG (2005) Symbiotic Hydra express a plant-like peroxidase gene during oogenesis. J Exp Biol 208:2157–2164

Nebel A, Bosch TCG (2012) Evolution of human longevity: lessons from Hydra (Editorial). Aging 4(11):730–731

Rahat M (1985) Competition between chlorellae in chimeric infections of Hydra viridis: the evolution of a stable symbiosis. J Cell Sci 77:87–92

Rahat M, Reich V (1983) A comparative study of tentacle regeneration and number in symbiotic and aposymbiotic Hydra viridis: effect of Zoochlorellae. J Exp Zool 227:63–68

Rahat M, Reich V (1984) Intracellular infection of aposymbiotic Hydra viridis by a foreign free-living Chlorella sp.: initiation of a stable symbiosis. J Cell Sci 65:265–277

Thorington G, Margulis L (1981) Hydra viridis: transfer of metabolites between Hydra and symbiotic algae. Biol Bull 160:175–188

Trembley A (1744) Mémoires, Pour Servir à l´Histoire d´un Genre de Polypes d´Eau Douce, à Bras en Frome de Cornes. Verbeek, Leiden

Wittlieb J, Khalturin K, Lohmann JU, Anton-Erxleben F, Bosch TCG (2006) Transgenic Hydra allow in vivo tracking of individual stem cells during morphogenesis. Proc Natl Acad Sci U S A 103(16):6208–6211

Corals

8

8.1 The Case of Reef Building Corals: A Complex Association Between Animal, Algal, and Bacterial Components

The reef-building corals that inhabit the warm, shallow waters of the tropics are perhaps the best-known example of a mutualism involving an animal, and in tropical marine environments, they are certainly one of the most significant. The enduring ability of corals (Scleractinia) and coral reefs to attract public interest and their ecological and economic importance stands in stark contrast to our ignorance of many basic aspects of their biology. In part, the lack of understanding is a consequence of the complexity of the association, but the net result is that we are seriously underequipped to understand why coral reefs everywhere are in serious decline and to implement policies that might at least slow this process.

Under the premise that the primary function of the immune system is to mediate host–microbe interactions, it follows that coexisting with a more complex microbiota requires a more functionally complex immune system, most likely with more components. In this context, it is interesting to compare the predicted innate immune repertoires of the coral *Acropora* with those of the sea anemone *Nematostella* and *Hydra*. This kind of comparison indicates that the numbers of pattern-recognition receptors, both for external molecules (Toll-like receptors, interleukin-1-like receptors, other transmembrane TIR proteins) and internal patterns (NBD proteins), are much greater in the coral than in the other cnidarians, and the diversity of domain combinations, particularly in the NBD proteins, is also greater. These characteristics are consistent with a requirement for more sophisticated control of host–microbial interactions in the coral than in other Cnidaria.

8.2 Attempts to Generalize About Coral–Microbe Interactions Are Complicated by the Evolutionary and Physiological Diversity of Corals

As is well known, shallow water reef-building corals are an association between a coral animal and a photosynthetic dinoflagellate belonging to the genus *Symbiodinium*. The association is an obligate one, at least as far as the coral animal is concerned—although these corals can survive for a short period after losing their partners, they die if an association with *Symbiodinium* is not reestablished in short order. It is much less well known that the Scleractinia, the order to which reef-building corals belong, contains almost as many species (more than 47 % of the 1490 recognized species; Cairns et al. 1999; Cairns 2007) that do not host (and do not require) *Symbiodinium*. These "azooxanthellate" corals frequently inhabit deep waters, where light levels are often very low and are mostly, but not exclusively, solitary—some azooxanthellate species are colonial and form massive deep-water reefs (Fig. 8.1). For example, the largest known *Lophelia* reef is the >100 km^2 Røst Reef off the Lofoten Islands in the Norwegian Sea. A small number of scleractinians are facultative hosts of *Symbiodinium*, for example, *Astrangia danae* and some species of *Oculina*.

While the presence or absence of photosymbiotic dinoflagellates must have major physiological consequences, substantial metabolic diversity clearly also exists across the range of zooxanthellate corals. The dinoflagellates that form symbioses with corals (*Symbiodinium* sp.) are an extremely diverse lineage, whose physiological characteristics are at least as diverse as those of the coral hosts (see, e.g., Klueter et al. 2015). Many corals acquire symbionts horizontally early in their life history and can form symbioses with a range of different *Symbiodinium* types, and in some cases the host can "switch" or "shuffle" symbiont strains, potentially resulting in the adjustment of holobiont physiology to environmental characteristics. While many species of reef-building corals are heavily dependent on their dinoflagellate partners for nutrition, heterotrophy is probably more important in others. For example, fungiid (mushroom) corals sometimes inhabit relatively turbid habitats where low light levels deter many other corals.

The taxonomy of the large number of recognized species of corals has traditionally been based on skeletal morphology but is often at odds with relationships based on DNA sequence data. For example, most classical families of zooxanthellate corals are not monophyletic on the basis of sequence data, and some of these families are split across the Complexa/Robusta divide—the deepest divergence among zooxanthellate corals. The Scleractinia have an ancient origin, most likely in the early Ordovician or late Cambrian (Stolarski et al. 2011), and much of the extant morphological diversity is represented in the mid-Triassic. On the other hand, some major speciation events are thought to be very recent—for example, the major diversification of *Acropora* sp. in the western Pacific most likely occurred 5–10 MYA. While, on classical grounds, this might be expected to have been sufficient time for the establishment of specific microbial communities, mass coral spawning events and introgression between many of these species may complicate or retard this process.

Fig. 8.1 Scleractinian corals colonize very different environments. (**a**) The *upper panel* shows a shallow water reef dominated by *Acropora millepora* in the Great Keppel Island system (central Great Barrier Reef) (From: van Oppen et al. (2015) *Peer J* DOI: 10.7717/peerj.1092. Photo credit: Alison Jones). (**b**) The *lower panel* shows a deep-water reef (Lopphavet, Norway) dominated by *Lophelia pertusa* (the white colonies to the right in the photo). (Image from: Geomar website (http://www.geomar.de/en/research/fb1/fb1-p-oz/research-topics/corals-reefs-climate-variability-on-short-time-scales/cold-water-corals/)

8.3 The Complexity of Coral Microbial Communities

The relationship between *Hydra* and its associated microbial communities is relatively well understood, and the *Hydra* system provides a useful framework for thinking about coral–microbial interactions. Whereas *Hydra* species have characteristic-associated bacterial communities (Franzenburg et al. 2013), the data for corals are equivocal on this point. Rohwer et al. (2002) provided evidence that each of three massive coral species (*Orbicella* (formerly *Montastraea*) *franksi*, *Diploria strigosa*, and *Porites astreoides*) contained distinct bacterial communities that were similar between distant locations (Panama and Bermuda) and over time (1 year). In apparent contradiction with this, significant differences were reported between bacterial communities associated with *Orbicella* (formerly *Montastaea*) *faveolata* from different sites near Tobago (Guppy and Bythell 2006) and communities associated with three different *Acropora* species on the Great Barrier Reef grouped according to sites rather than by species (Littman et al. 2009). One issue here is the complexity of coral taxonomy—and recent speciation in genera such as *Acropora*—compared to *Hydra*, where there are few species and divergences are deep. Approximately 150 *Acropora* species are currently recognized, whereas only a handful of *Hydra* species are recognized and divergences are significantly deeper in this genus. At this time, although there are consistent reports of seasonal variation and location effects, the consensus seems to be that different corals do harbor specific microbial populations, but this specificity is to some extent obscured by phylogenetic and technical complications. The microbial communities associated with corals are highly diverse and the application of different non-standard methods has contributed a great deal of "noise" into the search for patterns of association. Recent work (Ainsworth et al. 2015) implies that a core microbiome comprising a small number of taxa (seven) is common across three coral species (*Acropora granulosa* from the GBR, and two species from Hawaii—*Montipora capitata* and *Leptoseris* spp.) but also implies that very few taxa are universal within a species.

8.4　Where Are the Bacteria Located?

Interpreting the reported data for coral-associated microbes is complicated both by the structural complexity and heterogeneity of corals and also lack of a uniform methodology. Compared to a *Hydra* polyp, any coral colony presents a much greater diversity of micro-niches; the diversity in colony architectures across the Scleractinia is enormous, as is the within-colony heterogeneity. On top of this, there has been considerable variation in the methods used to isolate starting material for microbiome analyses, the net result of which is that studies vary with respect to how closely associated the bacteria are with the coral tissue.

As in other animals, bacteria are associated with coral epithelia, but in corals, bacteria are also present in coral mucous, others are associated with the skeleton, and some appear to be present inside coral cells. In the Caribbean coral *Montastraea cavernosa*, cyanobacteria are present within tissues (Fig. 8.2) and these have been implicated in nitrogen fixation within the association (Lesser et al. 2004). Although the taxa involved are diverse, nitrogen-fixing bacteria are a common component of coral microbiota, and the association between diazotrophs and corals is explored further below. In symbiotic corals, the zooxanthellae provide additional niches for microbes. FISH analysis suggests that, of the seven taxa comprising the core microbiome, the dominant two are closely associated with the symbionts—one (*Actinomycelates* sp.) appears to be within the dinoflagellate

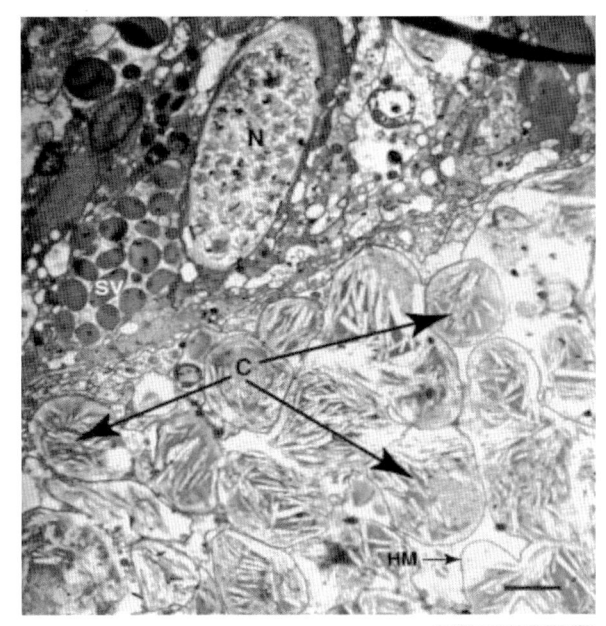

Fig. 8.2 Tissues of the Caribbean coral *Montastaea cavernosa* contain cyanobacteria (labeled as C in the figure), as revealed by electron microscopy of tissue sections. Similar structures are visible within acroporid corals from the Great Barrier Reef, and nitrogen-fixing bacteria of various types appear to be ubiquitous components of the microbial consortia associated with shallow water corals. Abbreviations: *N* nematocyst, *HM* host membrane, *SV* secretory vesicles. The scale bar represents 1 μm (From: Fiore et al. (2010))

cells, and the other (*Ralstonia* sp.) in the coral cells that contain *Symbiodinium* (Ainsworth et al. 2015).

Where data are available for several compartments, these imply that the bacterial communities of surrounding seawater and coral mucous are more similar to each other than is either to the coral tissue consortium. Mucous may have a "gatekeeper" function, providing a buffer zone between seawater and coral tissue, in which open warfare can be waged between bacteria based on antimicrobials produced both by the host and by resident bacteria, as well as quorum signaling between bacteria (Fig. 8.3).

A much more restricted set of species gain admission to coral tissue via the mucous gate.

8.5 Transmission Mode and Ontogeny

As in the case of *Hydra* species (see above), early life history stages of corals show highly diverse and variable bacterial associations that, in the case of broadcast-spawning species, are dynamic. Some evidence supports the idea that modes of

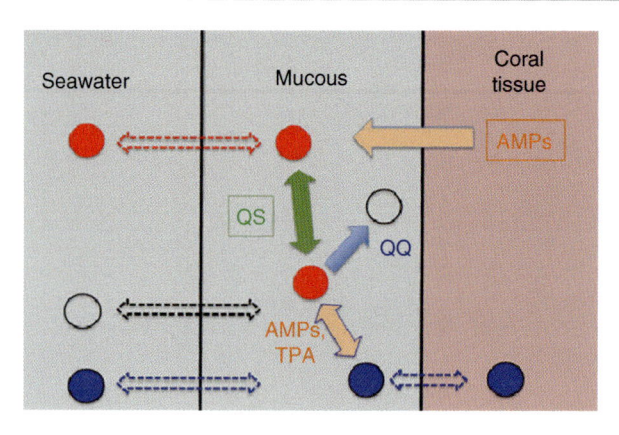

Fig. 8.3 Coral mucus may effectively act as a barrier between seawater and coral tissue, constraining potential pathogens in the case of healthy corals. Within the mucus layer, bacteria are subject to the actions of antimicrobial peptides (*AMPs*) produced by other bacteria and by the coral animal and bacterial metabolites such as tropodithietic acid (*TPA*). Bacterial communication via quorum sensing (*QS*) can be disrupted by quorum quenching (*QQ*) or blocking activities of other bacteria

transmission of bacteria may differ between brooding and broadcast-spawning corals, but only limited data are so far available so it may be premature to generalize at this stage. The presence of similar bacterial communities across early life history stages (from egg to planula larvae, but note that the consortia associated with adult corals differ; see below) of the brooding coral *Porites astreoides* (Sharp et al. 2012) is consistent with vertical transmission, whereas horizontal acquisition is implied by the fact that very different bacterial communities were associated with very early life history stages (oocytes, freshly released eggs) and planulae in the spawning coral *Pocillopora meandrina* (Apprill et al. 2012). The microbial communities associated with juvenile corals are typically more heterogeneous and less stable than are those of adult corals. Against this general backdrop, some consistent patterns can be seen, suggesting important functional roles for specific bacteria. For example, members of the *Roseobacter* clade are often major components of the bacterial communities associated with early life history stages of both brooding and broadcast-spawning corals but are often less well represented in adult microbiomes. In the case of *Acropora*, vertical transmission of Rhizobiales (alphaproteobacteria) implies that these nitrogen-fixing bacteria may play an important role in the holobiont (Lema et al. 2014).

8.6 Key Components of the Coral Microbiome

One clear pattern to emerge is that the microbial consortia associated with corals that form symbioses with *Symbiodinium* differ substantially from those associated with azooxanthellate corals, and corresponding differences have been documented for other classes of marine invertebrates (bivalves, octocorals, ascidians, sponges) with respect to whether or not photosynthetic symbionts (*Symbiodinium* or diatoms) are hosted (Bourne et al. 2013). The presence of the eukaryotic symbionts shapes the composition but not the diversity or complexity of the associated microbial communities. Of the two major classes of bacteria associated with reef invertebrates, gammaproteobacteria are often more abundant in photosymbiotic hosts, whereas they are outnumbered by alphaproteobacteria in non-photosymbiotic hosts (Bourne et al. 2013). However, it is clear that substantial variation exists in the relative proportions of alpha- and gammaproteobacteria even within the same coral species (Littman et al. 2009).

Members of a few bacterial genera, including *Roseobacter* "affiliates" and *Endozoicomonas/Spongiobacter*, are often major components of coral microbiomes, and many of these have properties in vitro that are consistent with beneficial effects in vivo, including nutritional or probiotic/antimicrobial properties.

The *Roseobacter* clade of alphaproteobacteria, which includes a number of genera frequently identified in association with cnidarians (*Ruegeria, Phaeobacter, Roseobacter*), are near ubiquitous in marine environments and important players in global sulfur and carbon cycling. Roseobacteria are generally mixotrophs capable of aerobic anoxygenic photosynthesis. Some strains are also apparently capable of anaerobic growth, using DMSO or nitrate as alternative terminal electron acceptors. Others produce DMS by degrading algal osmolytes. *Roseobacter* clade "affiliates" dominate the early life history stages of many corals but typically become less numerous later in development (for an example of this, in Fig. 8.4, compare the abundance of alphaproteobacteria in coral planulae and 1 week juveniles with the 12-month time point). The well-documented ability of many *Roseobacter* affiliates to produce the broadly active antimicrobial agent tropodithietic acid (TPA) may be the physiological basis of their association with corals. Also of relevance to the association with zooxanthellate corals is the fact that "symbiotic" relationships have been described between many *Roseobacter* clade members and dinoflagellates or other microalgae. For example, *Dinoroseobacter shibae* strains are symbionts of the toxic dinoflagellates *Pfiesteria, Alexandrium ostenfeldii*, and *Protoceratium reticulatum* as well as of other microalgae. At least one of these "symbiotic" associations is obligate: the association of *Ruegeria* TM1040 and the dinoflagellate *Pfiesteria piscicida. Pfiesteria* is unable to grow (and eventually dies) in the absence of *Ruegeria* TM1040, and the metabolic basis of the requirement is that the latter can synthesize vitamins B1 and B12, for which the dinoflagellate partner is auxotrophic. Roseobacteria have been shown to physically attach to dinoflagellates in culture, and there are hints that this may also be the case in coral tissue. How the bacteria benefit from association with dinoflagellates is unclear; however, DMSP is produced by many dinoflagellates and this compound is a known chemoattractant for some *Roseobacter* strains.

Endozoicomonas (*Spongiobacter*; gammaproteobacteria) and other Oceanospirillae have been identified in association with a wide range of marine invertebrates, including sponges, octocorals, and scleractinians. Oceanospirillae are dominant members of the bacterial populations associated with adult *Porites* sp.,

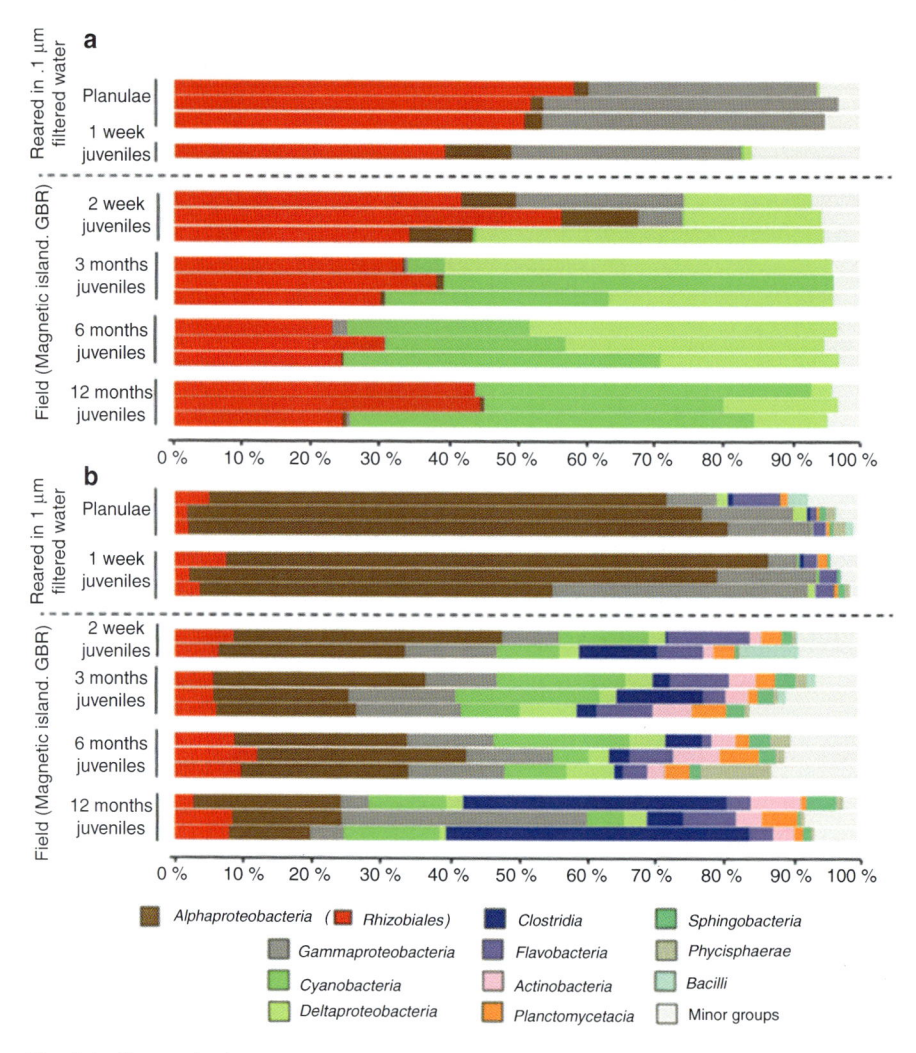

Fig. 8.4 Changes in the composition of bacterial consortia associated with *Acropora millepora* during ontogeny on the basis of (**a**, *upper*) nifH or (**b**, *lower*) partial 16S sequences. Note the early dominance of alphaproteobacteria and significant representation of Rhizobiales throughout ontogeny (From: Lema et al. 2014)

whereas *Roseobacter* and *Marinobacter* are more abundant in larvae. *Endozoicomonas* sp. are dominant in the tissues of adults of a wide range of anthozoans, including those with dinoflagellate photosymbionts, for example, *Stylophora pistillata*, as well as those without, for example the gorgonian *Eunicella verrucosa*. Many Oceanospirillae, including those in association with the coral *Acropora*, are capable of metabolizing **DMSP** to **DMS**.

8.7 Nitrogen-Fixing Bacteria Are Intimately Associated with Corals

The presence within tissues of the Caribbean coral *Montastraea cavernosa* of cyanobacteria capable of fixing nitrogen was reported more than 10 years ago (Lesser et al. 2004). Stable isotope ratio data indicate that the fixed nitrogen is primarily used by the resident *Symbiodinium* rather than by the host coral (Lesser et al. 2007); diurnal patterns of nitrogen fixation, with peaks during the evening and early morning, are negatively correlated with high tissue oxygen levels, consistent with the oxygen sensitivity of nitrogenase. The endosymbiotic nitrogen-fixing bacteria originally identified in *Montastraea* were described as coccoid cyanobacteria resembling *Synechococcus* or *Prochlorococcus* (Lesser et al. 2004), and bacteria with similar cellular ultrastructure have been described in *Acropora* spp. from the Great Barrier Reef (Kvennnefors and Roff 2009). However, more comprehensive PCR-based surveys of nitrogenase (nifH) gene sequences imply the presence of a diverse and variable range of nitrogen-fixing bacteria associated with both *Montastraea* and other corals. nifH sequence analyses imply that, whereas a variable range of diazotrophs are associated with coral mucous, the tissue resident community of nitrogen-fixing bacteria is stable (homogenous between locations) and specific for each of the three coral species, *Acropora millepora*, *A. muricata*, and *Pocillopora damicornis*, being dominated in each case by members of the Rhizobiales (Lema et al. 2012). Whereas big shifts occur during coral ontogeny in the overall composition of microbial communities, *Bradyrhizobium* sp. (Rhizobiales; see Fig. 8.4) are a dominant component of the diazotroph community associated with *Acropora millepora* from early stages to adults (Lema et al. 2014), which is consistent with a basic metabolic requirement. Shifts occur during ontogeny with respect to other diazotrophs, *Vibrio* sp. (gammaproteobacteria) being more abundant in planulae, whereas cyanobacteria become prevalent in late juveniles. Other studies suggest that different bacterial genera may dominate the diazotroph communities of other coral species; for example, *Vibrio*-like sequences accounted for the largest number of nifH reads in two Hawaiian species of *Montipora* (Olson et al. 2009). The caveat here, however, is the lack of uniformity in the methods applied; until methods are standardized, generalizations may be premature.

8.8 Probiotic Microbes and Antimicrobial Peptides

One idea prevalent in the literature is that some mucous-resident bacteria are able to prevent the growth of potential coral pathogens and thus benefit the host coral. While the idea of probiotic bacteria associated with coral mucous is an attractive one, to date the evidence for this is largely circumstantial. A number of studies have demonstrated that some bacteria isolated from coral mucous, particularly *Vibrio*, *Pseudoalteromonas*, and *Roseobacter* sp. (see, e.g., Shnit-Orland and Kushmaro 2009), have antimicrobial activity in vitro, but evidence for activity in coral mucous is lacking. As members of the *Roseobacter* clade are known to produce the antimicrobial compound tropodithietic acid (TPA), this compound may be responsible for the observed ability of isolates to prevent bacterial growth in vitro.

Whereas *Hydra* species produce a wide variety of AMPs, to date only one AMP has been characterized from any coral, the 40 AA residue damicornin peptide having similarity to vertebrate cysteine-rich defensins. The pattern of cysteine residues in the mature damicornin peptide is similar to those in an AMP identified in the jellyfish *Aurelia* and some channel-blocking toxins from sea anemones. The biologically active peptide is generated from a pre-proprotein, which contains a signal peptide and a dibasic cleavage site between the N-terminal acidic prodomain and the basic mature protein. This protein was identified in tissues of heat-stressed tissues of *Pocillopora damicornis* but, somewhat counterintuitively, its expression is repressed after exposure to the coral pathogen *Vibrio coralliilyticus*. Damicornin is nominally active both against gram-negative and gram-positive bacteria, as well as the fungus *Fusarium oxysporum*.

The paucity of literature on AMPs in corals should not be interpreted to mean that none are present; rather, the experimental work necessary for the identification of AMPs has not been done. Given the near ubiquitous production of AMPs by animals and the diverse repertoire of active molecules in *Hydra*, it would be surprising if corals were not also rich sources of proteins with antimicrobial activities.

8.9 Coral–Bacterial Interactions Modulate Local Climate Via Sulfur Metabolites

One intriguing aspect of coral/bacterial interactions involves the sulfur metabolite dimethylsulfoniopropionate (DMSP) and its breakdown product dimethylsulfide (DMS), both of which are key intermediates in global sulfur cycling (Fig. 8.5). The production of DMSP by marine algae has been extensively studied, and it has been estimated that diatoms and dinoflagellates, together with a few other groups of marine algae, collectively produce more than half of the biogenic sulfur released into the atmosphere. Coral reefs are known to be major sources of DMSP (Broadbent and Jones 2004). While most of the DMSP produced is used as a source of carbon and sulfur by bacteria, a significant proportion is converted to the volatile compound DMS though microbial activity. The latter compound is a "climate-active gas"—although there is some doubt about its impact on a global scale, DMS plays an important role in local (and probably regional) climate modification through its oxidation products inducing cloud formation, effectively lowering water temperatures and reducing photosynthesis on reefs (Vallina and Simó 2007).

Until recently it was believed that DMSP production by coral reefs was purely attributable to the dinoflagellate symbionts of corals and giant clams. However, it is now clear that the coral animal has the enzymatic machinery for DMSP production and that symbiont-free juvenile corals release significant amounts of DMSP (Raina et al. 2013). The most thoroughly investigated role of

8.9 Coral–Bacterial Interactions Modulate Local Climate Via Sulfur Metabolites

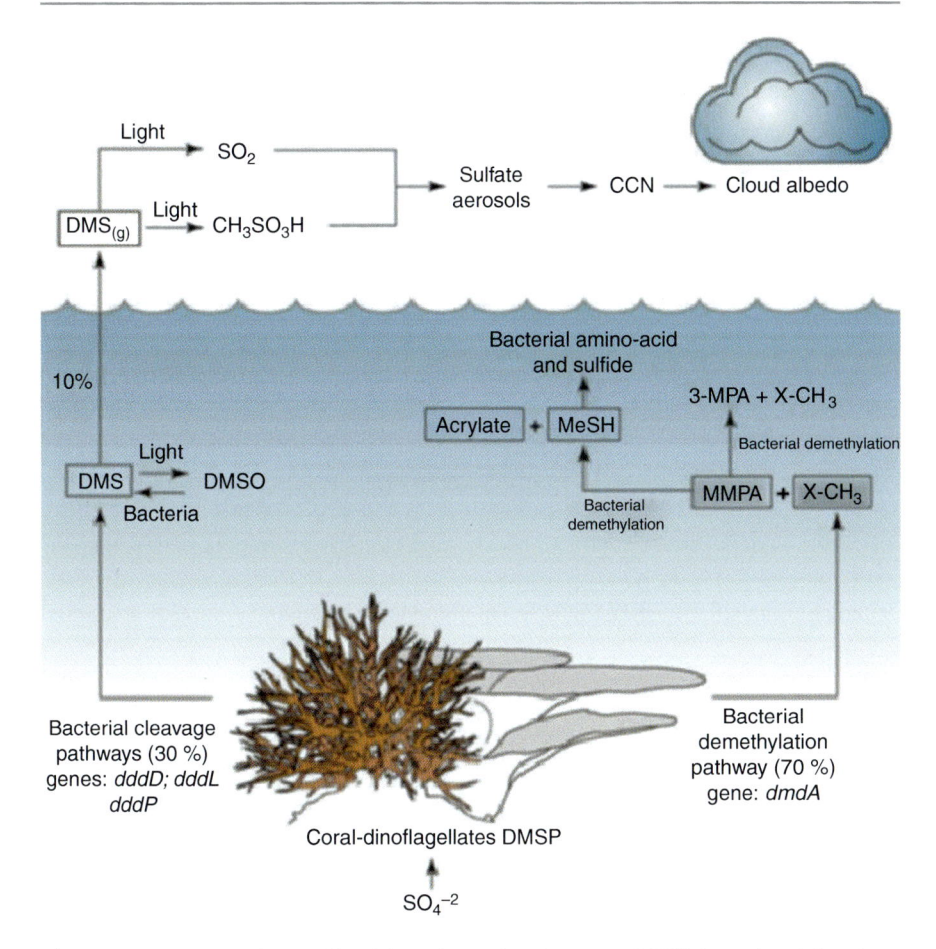

Fig. 8.5 DMSP production and breakdown in coral reef systems. DMSP is produced from inorganic sulfur in the coral–dinoflagellate association. A significant proportion of the DMSP produced is metabolized to DMS by coral-associated bacteria. DMS is volatile and can ultimately give rise to sulfate aerosols that alter cloud albedo (From: Raina et al. (2010))

DMSP is as an osmolyte, but it also has antioxidant and radical scavenging properties, so its function in corals is unclear. Production of DMSP may be a double-edged sword, however; DMSP production increases in heat-stressed corals, enabling the coral pathogen *Vibrio coralliilyticus* to identify colonies under stress (Garren et al. 2014).

Conclusion
1. The enormous physiological and evolutionary diversity of corals complicate unraveling general principles of coral–microbe interactions.
2. There is evidence for both species-restricted microbial consortia and environmental variation of consortia within a coral species.

3. Differences exist in transmission mode of bacteria; in some cases, the eggs contain bacteria, but in others bacteria are acquired from the environment. This may reflect brooding versus broadcast spawning reproduction.
4. *Roseobacter* clade bacteria (alphaproteobacteria) often dominate early life history stages of corals but become less numerous later. This interaction may be based on the ability of many *Roseobacter* affiliates to produce the general antimicrobial compound tropodithietic acid (TPA).
5. *Endozoicomonas* and its relatives (gammaproteobacteria) are often dominant members of the microbial consortia associated with adult corals and other anthozoans, irrespective of the presence/absence of photosymbionts.
6. Although each species has a characteristic profile of nitrogen-fixing bacteria, Rhizobiales are often represented in coral microbiomes, irrespective of life history stage or species.

References

Ainsworth TD et al (2015) The coral core microbiome identifies rare bacterial taxa as ubiquitous endosymbionts. ISME J 9(10):2261–74. doi:10.1038/ismej.2015.39

Apprill A et al (2012) Specificity of associations between bacteria and the coral *Pocillopora meandrina* during early development. Appl Environ Microbiol 78:7467–7475

Bourne D et al (2013) Changes in coral-associated microbial communities during a bleaching event. ISME J 2:350–363

Broadbent AD, Jones GB (2004) DMS and DMSP in mucus ropes, coral mucus, surface films and sediment pore waters from coral reefs of the Great Barrier Reef. Marine and Freshwater Research 55:849–855

Cairns SD (2007) Deep-water corals: an overview with special reference to diversity and distribution of deep-water scleractinian corals. Bull Mar Sci 81(3):311–322

Cairns SD, Hoeksema BW, van der Land J (1999) Appendix: list of extant stony corals. Atoll Res Bull 459:13–46

Fiore CL et al (2010) Nitrogen fixation and nitrogen transformations in marine symbioses. Trends Microbiol 18:455–463

Franzenburg S et al (2013) Distinct antimicrobial peptide expression determines host species-specific bacterial associations. Proc Natl Acad Sci U S A 110:E3730–E3738

Garren M et al (2014) A bacterial pathogen uses dimethylsulfoniopropionate as a cue to target heat-stressed corals. ISME J 8:999–1007

Guppy R, Bythell JC (2006) Environmental effects on bacterial diversity in the surface mucus layer of the reef coral *Montastaea faveolata*. Mar Ecol Prog Ser 328:133–142

Klueter A et al (2015) Taxonomic and environmental variation of metabolite profiles in marine dinoflagellates of the genus *Symbiodinium*. Metabolites 5:74–99

Kvennnefors ECF, Roff G (2009) Evidence of cyanobacteria-like endosymbionts in Acroporid corals from the Great Barrier Reef. Coral Reefs 28:547

Lema KA et al (2012) Corals form characteristic associations with symbiotic nitrogen-fixing bacteria. Appl Environ Microbiol 78:3136–3144

Lema KA et al (2014) Onset and establishment of diazotrophs and other bacterial associates in the early life-history stages of the coral *Acropora millepora*. Mol Ecol 23:4682–4695

Lesser MP et al (2004) Discovery of symbiotic nitrogen-fixing cyanobacteria in corals. Science 305:997–1000

Lesser MP et al (2007) Nitrogen fixation by symbiotic cyanobacteria provides a source of nitrogen for the scleractinian coral Montastaea cavernosa. Mar Ecol Prog Ser 346:143–152

Littman RA et al (2009) Diversities of coral-associated bacteria differ with location, but not species, for three acroporid corals on the Great Barrier Reef. FEMS Microbiol Ecol 68:152–163

Olson ND et al (2009) Diazotrophic bacteria associated with Hawaiian *Montipora* corals: diversity and abundance in correlation with symbiotic dinoflagellates. J Exp Mar Biol Ecol 371:140–146

van Oppen MJ, Lukoschek V, Berkelmans R, Peplow LM, Jones AM (2015) A population genetic assessment of coral recovery on highly disturbed reefs of the Keppel Island archipelago in the southern Great Barrier Reef. Peer J 3:e1092

Raina J-B et al (2010) Do the organic sulfur compounds DMSP and DMS drive coral microbial associations? Trends Microbiol 18:101–108

Raina J-B et al (2013) DMSP biosynthesis by an animal and its role in coral thermal stress response. Nature 502:677–680

Rohwer F et al (2002) Diversity and distribution of coral-associated bacteria. Mar Ecol Prog Ser 243:1–10

Sharp KH et al (2012) Diversity and dynamics of bacterial communities in early life history stages of the Caribbean coral *Porites astreoides*. ISME J 6:790–801

Shnit-Orland M, Kushmaro A (2009) Coral mucus-associated bacteria: a possible first line of defense. FEMS Microbiol Ecol 67:371–380

Stolarski J et al (2011) The ancient evolutionary origins of Scleractinia revealed by azooxanthellate corals. BMC Evol Biol 11:316

Vallina SM, Simó R (2007) Strong relationship between DMS and the solar radiation dose over the global surface ocean. Science 315:506–508

Bleaching as an Obvious Dysbiosis in Corals

9

For many shallow water reef-building Scleractinia, the association between the coral and its resident dinoflagellates is relatively unstable, collapsing after prolonged exposure to high seawater temperatures. The result of this collapse is departure of the dinoflagellate from the coral host, but at this time it is not clear which partner initiates the separation or what the trigger is. It is clear that thermal anomalies can cause bleaching and that this effect is exacerbated by high UV exposure. Whatever the trigger, the net result is "bleaching" of the coral colony; corals which have lost their dinoflagellate symbionts often have no pigmentation and, as the skeleton can usually be clearly seen, they appear as white ghosts of their former selves (Fig. 9.1). The bleached state is metastable—the coral can quickly recover if a compatible *Symbiodinium* strain can either be taken up from an external source or can grow back from a low background—but, if the coral fails to reestablish the photosymbiosis within a relatively short time, death follows. It has been suggested that coral bleaching might have adaptive value, in that it might enable "switching" or "shuffling" to a strain of *Symbiodinium* whose physiology might be more compatible to the host under changed environmental conditions. A major concern at present, however, is that as seawater temperatures increase, the threshold for bleaching will be reached more often, so frequent mass coral mortality will occur, perhaps resulting in complete demise of coral reefs.

The summary above is essentially how the "coral reef crisis" is often presented to the informed public, with putative mechanisms thrown in to make the story sound more scientifically respectable. While intrinsically attractive, this narrative is an oversimplification that does not adequately consider the physiological variability of corals and ignores inconvenient facts. In fact, enormous variation exists both between and within species in beaching sensitivity. Because *Acropora* species dominate most Indo-Pacific reefs and are particularly susceptible to bleaching, most attention has focused on these. However, other genera such as *Porites* or *Goniastrea* are far more tolerant of thermal stress. In addition, species show variation in sensitivity across their distribution ranges, which often encompass very different thermal

Fig. 9.1 Example of bleached coral (*white*) with healthy coral in the Mariana Islands, Guam (Photo credit: David Burdick, NOAA. Read more: http://marinesciencetoday.com/2012/10/16/more-algae-means-corals-have-more-to-lose/#ixzz3iHWGYpuH)

environments. For example, the distribution ranges of several *Acropora* species include the Great Barrier Reef and the Red Sea/Persian Gulf; populations from the Red Sea thrive under temperature conditions under which GBR populations would undergo mass mortality. Some of the variation in thermal tolerance is associated with hosting specific *Symbiodinium* strains, but there is evidence that both host and symbiont genotypes are involved and that acclimatization can also occur.

The most widely accepted hypotheses to account for coral bleaching are built around heat-induced damage to photosynthetic systems as triggers for the collapse; either direct damage to components of PSII or over-reduction of electron carriers, ultimately resulting in oxidative damage to photosynthetic proteins. However, the view that bleaching occurs as a consequence of damage to the photosynthesis machinery is difficult to reconcile with the observation that dinoflagellates expelled from heat-stressed corals are photosynthetically competent (Ralph et al. 2001).

Bleaching is triggered not only by exposure to elevated temperatures and UV but also to a wide range of other stressors in vitro or on reefs, including exposure to suboptimal temperatures, cyanide, low salinity, and turbidity. Moreover, it has been suggested that microbial pathogens or changes more broadly in the microbial consortia associated with the coral may be the real triggers for bleaching (see below). To complicate the matter further, "bleaching" is an ambiguous term—it is the most obvious morphological expression of a range of phenomena that includes not only expulsion of symbionts but also their breakdown within the host, the detachment of large numbers of whole coral cells containing symbionts, or simple pigment destruction (Douglas 2003).

9.1 The Complex Relationship Between Stress Sensitivity and the Transmission Mode and Diversity of Symbionts

As noted above, coral holobionts vary enormously in their sensitivity to stress, with *Acropora* spp. being notoriously sensitive, whereas, for example, members of the genus *Porites* are typically much less so. These two genera differ in some other important characteristics, including mode of symbiont transmission, with *Porites* spp. having maternal transmission of *Symbiodinium*, whereas *Acropora* spp. acquire symbionts post-settlement. It has been suggested that the two traits are linked—perhaps corals that use maternal transmission are intrinsically less sensitive to stress, the measure of stress sensitivity being bleaching susceptibility in response to thermal stress. Maternal transmission of symbionts could be viewed as reflecting a tighter integration of the two partners, perhaps reflecting a higher level of coevolution or co-adaptation; hence the symbiosis may be less prone to collapse under stress.

While superficially attractive, the idea that maternal transmission brings with it greater stress resistance does not stand up to scrutiny. While *Porites* spp. in general are less prone to bleaching than are *Acropora* spp., the pattern breaks down upon consideration of a broader range of corals. *Montipora* is a genus closely related to *Acropora* but differs from it in that transmission of symbionts is typically maternal, and at least some *Montipora* spp. undergo massive bleaching in response to stress. However, the nature of stress sensitivity differs between *Acropora* and *Montipora*—the latter being more susceptible to low salinity and sediment than to thermal stress. *Stylophora pistillata*, which is frequently employed for in vitro studies of coral biology, transmits maternally but is as prone to bleach under thermal stress as is *Acropora*.

The examples provided above indicate that the relationship between symbiont transmission mode and stress sensitivity is not a simple one—maternal transmission of symbionts does not necessarily bring with it greater stress tolerance. Instead of ensuring transmission of a coevolved symbiont, perhaps maternal inheritance is a mechanism that provides the offspring with a selection of options—different strains of zooxanthellae from which specific types can be selected on the basis of environmental variables.

A related idea is that stress sensitivity is reflected in symbiont diversity in the holobiont; here, mode of transmission is ignored, but the presence of a homogenous symbiont community in host samples from different environments is seen as characteristic of stress-tolerant coral species (e.g., *Porites* spp.), whereas more heterogeneous symbionts are found in more stress-sensitive corals (e.g., *Acropora* spp. and *Pocillopora* spp.). In the literature, these are referred to as "specifists" (low symbiont diversity) and "generalists" (high symbiont diversity).

To some extent, this is a tautological argument—"generalist" corals by definition can "shuffle" or "switch" and must lose symbionts to do so, and bleaching is a simple means of achieving this. Moreover, estimating symbiont diversity is a complex and arbitrary process, and Andrew Baker's group has elegantly demonstrated extensive cryptic diversity in symbiont types associated even with "specifist" corals such as *Porites* spp. (Silverstein et al. 2012). Symbiont diversity appears to be much greater than has been assumed, and conclusions about symbiont specificity and stress tolerance are not justified on the basis of the data presently available.

9.2 Do Bacteria Cause Coral Bleaching?

Eugene Rosenberg's work in the early 2000s provided a completely novel perspective on coral bleaching which has proved to be highly controversial.

The annual beaching of the scleractinian coral *Oculina patagonica* in the eastern Mediterranean, which occurred when seawater temperatures rose during the late summers of the period 1995–2003, appeared to be a clear-cut example of bacterial infection driving coral bleaching and thus promised a better understanding of the underlying molecular mechanisms.

O. patagonica is a fairly hardy coral, frequently inhabiting rock pools where temperatures may exceed 40 °C and salinity 5 %, but bleaching events in the late 1990s and early 2000s lagged behind the average sea surface temperature. In surveys of bacteria associated with *O. patagonica*, the presence of one specific strain of *Vibrio*—*V. shiloi* (AK1)—was directly correlated with bleaching state; *V. shiloi* AK1 was present in 28 of 28 bleached colonies and absent from 24 of 24 nonbleached colonies. *V. shiloi* AK1 satisfied all of Koch's postulates, including the ability to cause bleaching (120 bacteria/ml caused 100 % bleaching in 20 days at 29 °C) and was thus considered to be the causative agent (reviewed in Rosenberg and Falkovitz 2004). Further research clarified the mechanisms involved in infection and pathogenesis (reviewed in Bourne et al. 2009), including chemotaxis and adhesion to a component of coral mucus, invasion of and differentiation in epidermal cells, intracellular multiplication, and production of toxins that inhibit photosynthesis and a self-protective superoxide dismutase (SOD). In this model, temperature acted as an environmental trigger to outbreaks, altering virulence via increasing adhesion and the production of toxin and SOD.

While the *O. patagonica*/*V. shiloi* AK1 system promised deeper insights into both the mechanism of bleaching and coral–bacterial interactions, after 2003 *V. shiloi* was no longer consistently associated with bleaching in *O. patagonica*, so this entire line of investigation essentially dried up. The sudden change in what appeared to be a strong association could be due to loss of a virulence factor on the part of the bacteria or acquisition of resistance or tolerance on the part of the coral and has some parallels in the case of another cnidarian/pathogen interaction, that of the fungal pathogen *Aspergillus sydowii* with gorgonians in the Caribbean.

Regardless, that bleaching still regularly occurs and is not associated with *V. shiloi* indicates that there are other causes of bleaching.

9.3 Coral Disease and the Significance of Opportunistic Pathogens

A second documented example of coral bleaching apparently being triggered by bacterial infection is the case of *Pocillopora damicornis* and *Vibrio coralliilyticus*, which is also a pathogen of a number of other marine species, including bivalve mollusks, crustaceans, and fish. Whereas *O. patagonica* is a temperate coral that does not build reefs, *P. damicornis* is a tropical reef builder. *V. coralliilyticus* was first isolated from a bleached colony of *P. damicornis* from a reef near Zanzibar in the Indian Ocean. The same bacterial species was subsequently identified from a number of bleached *P. damicornis* colonies from reefs in Eilat in the Red Sea but was not detected in healthy colonies. The bleaching pathology associated with *V. coralliilyticus* is, however, complex and temperature dependent. Whereas a number of *V. coralliilyticus* strains are coral pathogens, only some strains (including the type strain BAA-450) cause bleaching. *V. coralliilyticus* BAA-450 is avirulent below 22 °C, whereas infection in the range 24–26 °C causes bleaching, but rapid tissue lysis occurs following infection at 27–29 °C, apparently as a consequence of expression of a metalloproteinase at temperatures above 26 °C. Other strains of *V. coralliilyticus* are implicated in other coral diseases. Strain OCN008 is associated with acute *Montipora* white syndrome (Ushijima et al. 2014).

V. coralliilyticus is one of only a handful of coral pathogens that fulfill Koch's postulates (Ushijima et al. 2014), the others being *V. shiloi* (but see above), *V. owensii*, *Aurantimonas coralicida*, and the human pathogen *Serratia marcescens*.

9.4 Changes in Coral-Associated Microbial Consortia Under Stress

While the specific cases outlined above implicate single pathogens as causative agents in coral disease, the emerging picture is that a wide variety of organisms are potentially opportunistic pathogens of corals; hence pathogenesis often manifests itself by shifts in abundance of coral-associated bacteria rather than presence/absence of a specific pathogen. The essence of the coral probiotic hypothesis (Reshef et al. 2006) is that the health of the coral animal depends on a "healthy" microbiota—the bacterial symbionts and their interactions may effectively exclude potentially pathogenic bacteria. Increased abundance of *Vibrio*-affiliated sequences is frequently associated with pathogenic conditions.

While the literature is consistent with the idea that stress causes major changes in the microbial consortia associated with particular coral species and the link between such shifts and the diseased state has often been made, our poor understanding of the "normal" microbiota (see above) and coral–microbe interactions in the "healthy" state mean that all such claims should be examined critically. Nevertheless, there is evidence for major changes in the associated microbial community in corals during a natural bleaching episode. During a mass bleaching event in the (southern hemisphere) summer of 2002, the bacterial community associated with *A. millepora* colonies that bleached shifted toward *Vibrio* affiliates, whereas *Spongiobacter/Endozoicomonas* affiliates were associated with non-bleached samples (Bourne et al. 2008). In this case, the shift toward *Vibrio*-dominated communities was detected prior to bleaching being visually detectable.

Decreased pH (increased pCO_2) caused clear changes in the microbiota associated with the coral *Acropora millepora* (Webster et al. 2013). After 6 weeks of exposure to seawater equilibrated with elevated CO_2, two alphaproteobacteria were lost and both loss and gain of specific gammaproteobacteria had occurred. The gamma species gained had sequence similarity to organisms previously associated with stressed and diseased corals, suggesting that at higher pCO_2, a shift had occurred toward more opportunistic bacteria.

Disease outbreaks are often associated with thermal anomalies or other stressors; typically outbreaks of disease follow periods of abnormally high sea surface temperatures, bleaching events occurring simultaneously with or following disease symptoms. Note that there are exceptions to this—for example, the main drivers of "atramentous necrosis" (AtN) outbreaks on *Montipora aequituberculata* are other stressors—low salinity and suspended matter being more significantly associated with AtN than temperature (Haapkyla et al. 2011).

It is entirely feasible that bleaching is directly triggered either by pathogenic microbes or shifts in the associated microbiota.

9.5   *Symbiodinium* as a Recent Intruder on Preexisting Coral–Bacterial Mutualisms

On evolutionary time scales, the coral–*Symbiodinium* symbiosis is relatively recent. "Relatively" is an important qualifier here; the relationship has been in place for 50–65 million years (Tchernov et al. 2004; Pochon et al. 2006). Coral fossils from the mid-Triassic (240 MYA) have traditionally been regarded as marking the origins of the Scleractinia, but the inclusion of data for a representative range of deep-water (azooxanthellate) Scleractinia has pushed coral origins significantly deeper—most likely in the Cambrian (Stolarski et al. 2011). Like extant azooxanthellate corals, the ancestors of the corals responsible for building the shallow water reefs of modern times were devoid of dinoflagellate symbionts.

The symbiosis between coral animal and dinoflagellate has been remarkably successful, but it is likely that this particular photosynthetic symbiont was recently added into an already complex system—the ancestor(s) of these modern reef builders had lived in close association/symbiosis with complex microbial consortia for hundreds of millions of years, so *Symbiodinium* could be viewed as an intruder into the happy household. The success of this association implies that *Symbiodinium* brought considerable advantages with it when the symbiosis was first established. The mismatch in timing referred to above implies that, prior to hosting *Symbiodinium*, shallow water corals hosted other photosymbionts, so corals had presumably already solved some of the problems of entering the relationship. In exchange for the freedom from the shackles of reliance on heterotrophy which hosting photosymbionts brought, one obvious cost was that oxygen tensions in tissues oscillate massively by comparison with their non-symbiotic relatives. Oxygenic photosynthesis by day will make coral tissue hyperoxic, to a point that oxygen radicals may cause tissue damage, whereas algal consumption of oxygen during darkness potentially leads to tissue hypoxia. There is evidence that modern zooxanthellate corals have evolved mechanisms in response to these challenges; in *Acropora millepora*, transcription of a suite of molecular chaperones is hardwired to the circadian clock, so that "damage control" proteins are produced in preparation for, rather than in response to, the challenges of tissue hyperoxia

(Levy et al. 2011). The response to tissue hypoxia, on the other hand, is based on the oxygen-sensitive hypoxia-inducible factor (HIF) transcription factor system, which is highly conserved across the Metazoa.

Although it may have recently displaced an older kind of photosymbiont, the intrusion of *Symbiodinium* on preexisting relationships between corals and bacteria may have brought with it not only metabolic challenges such as management of nutrient exchange processes but also biological hazards. Dinoflagellates frequently have intimate relationships of their own, often (as indicated above) with *Roseobacter*-related bacteria. In all likelihood, the ancestor of *Symbiodinium* had an associated microbial consortium, possibly including *Roseobacter* relatives, which came with it when it first explored opportunities for symbiosis with ancestral corals. Although the coral/*Symbiodinium*/bacteria relationship is generally conceptualized in terms of three interacting components, it may therefore be more appropriate to think of it in terms of four, so as to make clear that both host and photosynthetic symbiont bring their own bacterial complements to the association. Naturally this increased the complexity of interactions and provided new grounds for conflicts. The biochemistry of *Symbiodinium*–bacterial symbioses or commensalism is completely unexplored territory, but it is relevant to note that although it is thought that most or all *Symbiodinium* strains can be cultured ex host, optimal growth in vitro often requires the presence of specific bacteria. The genome of "clade B1" *Symbiodinium* (*S. minutum*) revealed the presence of a closely associated alphaproteobacterium, *Parvibaculum lavamentivorans*, a member of the order Rhizobiales (Shoguchi et al. 2012). Although there appear to be no consistent relationships between the microbiota of juvenile corals (*A. millepora* and *A. tenuis*) associated with the clade C1 or D strains of *Symbiodinium* (Littman et al. 2009), this is perhaps not surprising given the microbiome is highly dynamic and diverse at this life cycle stage. Given the immense metabolic diversity of *Symbiodinium* strains, it is hard to imagine that their contribution to the holobiont-associated microbial consortia does not differ. At the moment, the impact of *Symbiodinium*-associated bacteria on the microbial consortia of the holobiont can only be guessed at; however, it is not unreasonable to speculate that the bleaching response is actually directed against the bacteria associated with *Symbiodinium*. Perhaps stressors result in changes in the abundance of such bacteria (possibly *Roseobacter* affiliates) to the extent that they challenge the survival of the coral host. Alternatively, if *Symbiodinium* has managed to negotiate entry to coral cells by making use of a system by which the host recognizes bacteria, *Symbiodinium* could be collateral damage in a host response to an imbalance in the holobiont bacterial consortium.

9.6   Coda: Are Coral Reefs Doomed?

For quite some time, there have been predictions about the imminent demise of coral reefs, the concerns centering on two climate change stressors, rising seawater temperature and decreasing ocean pH. Are these concerns warranted, and should temperature and CO_2 be the major concerns?

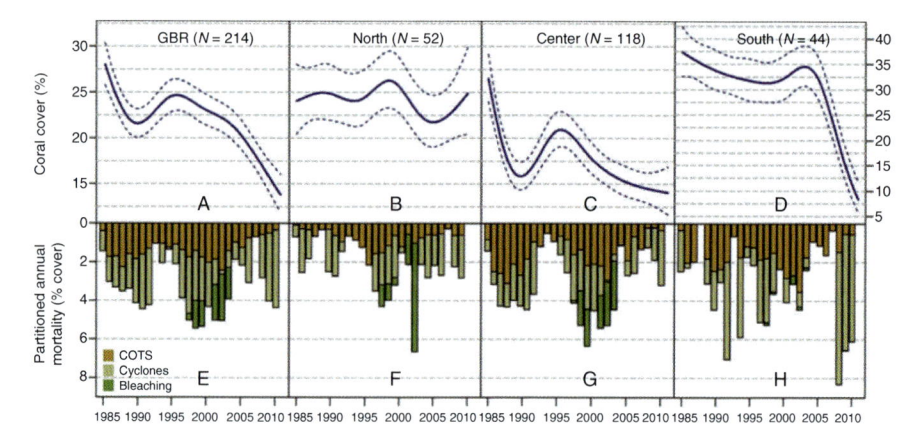

Fig. 9.2 Loss of coral cover (**a–d**) and annual mortality in the period 1985–2012 due to crown of thorns starfish outbreaks, cyclones, and bleaching (**e–h**) for the GBR as a whole and for the northern, central, and southern regions individually (*N* number of reefs). The *solid blue lines* in (**a–d**) represent estimated means (±2 SEs) for each trend. In (**e–h**), the composite bars represent mean coral mortality for each year, and the sub-bars indicate the relative mortality due to crown of thorns outbreaks, cyclones, and bleaching from De'ath et al. (2012)

There is irrefutable evidence for large-scale degradation of coral reefs globally. Estimates vary, but at least 50 % of reefs worldwide are highly degraded and most of the remainder at risk. Losses affect not only reefs close to underdeveloped nations but also those generally regarded to be in good shape, such as the Great Barrier Reef (GBR) in Australia. Despite a level of protection considered to be world's best practice, over 50 % of coral cover has been lost in the GBR over the period 1985–2012 (De'ath et al. 2012). Most of the Caribbean reefs are even more severely degraded, so there are genuine grounds for concern about the fate of coral reefs. However, it is informative to look at the causes of coral cover loss.

Despite expectations, in the case of the GBR, the main culprits have been cyclones (typhoons; 48 %) and outbreaks of the voracious crown-of-thorns starfish (42 %), with bleaching events accounting for around 10 % of loss during the 27-year period of the survey (Fig. 9.2). Although the predictions are that extreme weather events will increase in frequency as a result of climate change, the direct impact of temperature on coral reefs appears to be much less significant than are losses due to crown-of-thorns outbreaks. Water quality issues appear to be important in the case of starfish epidemics (Brodie et al. 2005), but the links remain elusive in part due to the long time scales involved.

9.6.1 The Geological Perspective: The Persistence of Coral Reefs

Global atmospheric CO_2 recently reached 400 ppm, against estimates of around 350 ppm for preindustrial times, causing ocean pH to decrease by about 0.1 unit since the 1950s. Under "business as usual" scenarios, atmospheric CO_2 is predicted

9.6 Coda: Are Coral Reefs Doomed?

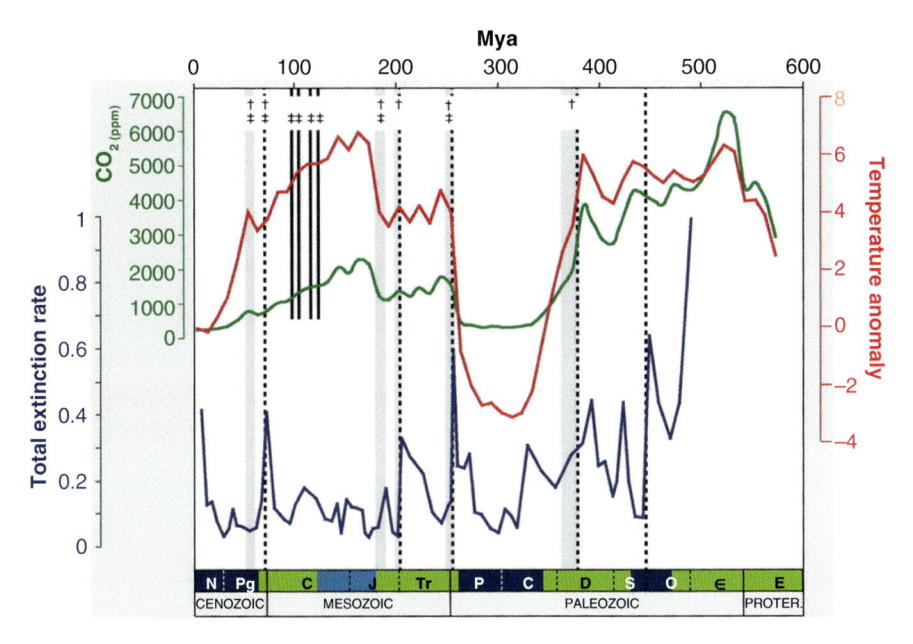

Fig. 9.3 Physical and biological changes affecting reefs through geological time. The green and blue colors on the x-axis reflect greenhouse and icehouse periods respectively, and the vertical broken lines indicate global mass extinction events. Vertical gray bars indicate reefs crises, and the shorter vertical black bars ocean acidification events; note that these latter occurred on relatively long time scales and are not associated with reef crises (From: Pandolfi et al. 2011)

to rise to perhaps twice that level by the end of the century, perhaps pushing ocean pH down by another 0.2–0.3 units. Higher atmospheric CO_2 is also causing increases in sea surface temperatures, the oceans having warmed around 0.6 °C in the last century, with the IPCC predicting further increases of perhaps 3 °C by 2100. Do these changes imply that we are seeing the end days of the Scleractinia as a lineage or of extensive coral reef systems? While outcomes of this sort have been predicted by some for decades, consideration of the evolutionary history of coral reefs and the Scleractinia suggest caution be exercised in extrapolating from short-term laboratory studies to the real world in real time. In fact the focus on ocean acidification and thermal stress could be viewed as a distraction from the main causes of loss of coral cover, some of which are certainly anthropogenic in origin and about which action might be more achievable in the short term—if even more politically challenging. An example is improvement of water quality, with its presumed relationship to crown-of-thorns outbreaks.

Coral reefs have been a feature of the tropical shallows for hundreds of millions of years, which at times have seen atmospheric CO_2 more than 20-fold higher than preindustrial levels and temperatures more than 7 °C higher than present-day levels (Fig. 9.3). So should we worry much about the future of corals? Although history suggests that the coral lineage will survive, there are real grounds for concern given the present rate of change. On geological time scales, atmospheric CO_2 (and thus ocean pH) is essentially uncoupled from seawater mineral (calcium carbonate) chemistry as a consequence of slow negative geochemical feedbacks (for example, increased rates of weathering of rocks under higher atmospheric CO_2) that result in increased ocean alkalinity. However, rapid increases in atmospheric CO_2 can exceed the capacity of feedback systems to compensate, and the predictions are that it will take thousands of years for the oceans to compensate for atmospheric changes that are currently being imposed on a scale of decades.

The major concern then is the rate, rather than the magnitude, of changes in temperature and atmospheric CO_2. The fossil record indicates that coral reefs have been subject to five "major crises"—times when global biodiversity loss or cessation of reef growth has occurred—and the most recent four of these have occurred when rapid decreases in ocean pH have occurred concurrently with global warming (Pandolfi et al. 2011). Although this scenario should sound familiar, an interesting exception to this general pattern occurred during the most recent reef crisis. During the Paleocene–Eocene thermal maximum (55.8 MYA), when rates of temperature and CO_2 change were comparable to those presently occurring, at least one reef system was largely unaffected. Other reef settings do appear to have suffered—sudden changes from coral- to foraminifera-dominated communities have been attributed to this thermal maximum—but the fact that at least one was not implies that reef systems may be more robust to rapid changes in temperature and pH than is generally assumed.

Over the last 20,000 years, reef systems have shown continuous growth and expansion despite a number of times facing rapid sea surface temperature (SST) increases, but this was against a background of lower SST than at present and relatively stable atmospheric CO_2 of around 330 ppm or lower, so optimism should be tempered by the fact that this apparent precedent is not strictly comparable with the present situation.

9.6.2 Impacts of Ocean Acidification on Corals

The main concern about the impact of OA on coral reefs is centered on the extent to which it will impact the calcification process—the deposition of calcium carbonate as aragonite in the coral skeleton. As ocean pH decreases, equilibria shift, resulting in decreased availability of carbonate ions, which are the substrate for the calcification process. Calcification becomes more energetically costly with decreasing pH, and early attempts to predict how OA would impact corals suggested that at around 580 ppm CO_2, coral reefs might start to dissolve rather than deposit calcium carbonate (Silverman et al. 2009). More refined estimates, based on a wide range of approaches, imply that although OA responses are consistently negative in impact, large differences in sensitivity exist between coral species, that responses are nonlinear, and that increased heterotrophy could at least partially compensate for the increased energetic cost of calcification under more acid conditions. Of course the other side of this rationale is that starved or stressed corals will fare even less well under OA, but in the short to medium term at least, OA per se is unlikely to drive massive losses of coral cover.

9.6.3 What About the Direct Impact of Thermal Stress or Elevated CO$_2$ on Corals?

To a great extent, what we know about the impact of thermal stress on corals is based on short-term laboratory experiments. In general, these support a direct link between oxidative stress, damage to some critical photosynthetic function, and bleaching. In laboratory settings, 33 °C is often cited as the critical temperature, prolonged exposure to this triggering bleaching in *Acropora* spp. and other corals. However, real-world patterns of coral bleaching do not always conform to predictions based on laboratory studies. As indicated earlier, coral thermal tolerance on reefs is strongly influenced by their previous experience; *Acropora* spp. on reefs in the Red Sea are often exposed for prolonged periods to 33 °C or higher, which would result in mortality of GBR colonies of the same species, and thermal tolerances differ along the barrier reef in line with "normal" temperature regimes in those regions. Corals are clearly adapted to their immediate thermal environments, and sufficient variation exists within many species to enable growth in very different temperature regimes. There is also convincing evidence that corals which have experienced sublethal stress acquire greater tolerance, either by host acclimation or adjusting the *Symbiodinium* repertoire, and the microbial consortia may also play important roles in this process.

9.6.4 Can Corals Evolve Fast Enough to Keep Pace with the Rate of Climate Change?

Recent work implies that at least some corals may be able to acclimate to more challenging (higher) temperatures remarkably quickly, and the same may also be true of elevated pCO$_2$. With the application of genomics-based technologies, the molecular processes underlying the acclimation process in corals, which may have an epigenetic basis, are being dissected. At least some coral species also contain remarkable levels of allelic variation, and introgression could permit beneficial alleles to spread between species—but the compelling issue is that of time.

Whereas generation times for *Drosophila* species, which have been demonstrated to adapt quickly to temperature increases, are measured in weeks, for corals it is more appropriate to think in terms of years. Given this fact, is it reasonable to expect a sessile marine animal to evolve the ability to tolerate more challenging temperature and pH regimes over the next few decades?

Although there are questions about the extent of acclimation that a single coral genotype may be able to achieve, the flexibility of the coral holobiont and the much shorter generation times of both *Symbiodinium* and bacterial components of the association presumably confer evolutionary advantages. Given the remarkable longevity of corals as a lineage and their spectacular evolutionary success in the past, the demise of reefs (and of the Great Barrier Reef in particular) would be a dramatic indictment of the real costs of economic progress.

> **Conclusion**
> 1. Coral bleaching is a visually obvious outcome that may reflect a number of quite different mechanisms that include not only expulsion of zooxanthellae but also their destruction within the coral host. The primary mechanisms of bleaching may differ between coral groups and different stressors may trigger different bleaching mechanisms.
> 2. The relationship between stress, disease, and bleaching is complex. Stress shifts the associated microbial communities in ways that probably facilitate disease but may also trigger bleaching.
> 3. Very few coral pathogens satisfy Koch's postulates; bacteria that are present in the consortia associated with "healthy" corals are opportunistic pathogens, but how the transition to pathogenicity is achieved is unclear.
> 4. *Vibrio coralliilyticus* strains are pathogens of many corals over a wide geographic range. Some other *Vibrio* spp. are either causative or opportunistic pathogens of corals, thus this group are particularly significant for coral health.
> 5. The relationship between *Symbiodinium* and corals is relatively recent, and its establishment most likely involved big changes in the microbial communities associated with the host. The bleaching response may reflect imbalances in those components of the microbiota that came with the symbiont.

References

Brodie J et al (2005) Are increased nutrient inputs responsible for more outbreaks of crown of thorns starfish – an appraisal of the evidence. Mar Pollut Bull 51:266–278

Bourne D et al (2008) Changes in coral-associated microbial communities during a bleaching event. ISME J 2:350–363

Bourne DG et al (2009) Microbial disease and the coral holobiont. Trends Microbiol 17:554–562

De'ath G et al (2012) The 27-year decline of coral cover on the Great Barrier Reef and its causes. Proc Natl Acad Sci U S A 109:17995–17999

Douglas AE (2003) Coral bleaching – how and why? Mar Pollut Bull 46:385–392

Haapkyla J et al (2011) Seasonal rainfall and runoff promote coral disease on an inshore reef. PLoS One 6:e16893

Levy O et al (2011) Complex diel cycles of gene expression in coral-algal symbiosis. Science 331:175

Littman RA et al (2009) Diversities of coral-associated bacteria differ with location, but not species, for three acroporid corals on the Great Barrier Reef. FEMS Microbiol Ecol 68:152–163

Pandolfi JM et al (2011) Projecting coral reef futures under global warming and ocean acidification. Science 333:418–422

Pochon X et al (2006) Molecular phylogeny, evolutionary rates and divergence timing of the symbiotic dinoflagellate genus *Symbiodinium*. Mol Phylogenet Evol 38:20–30

Ralph PJ et al (2001) Zooxanthellae expelled from bleached corals at 33oC are photosynthetically competent. Mar Ecol Prog Ser 220:163–168

Reshef L et al (2006) The coral probiotic hypothesis. Environ Microbiol 8:2068–2073

Rosenberg E, Falkovitz L (2004) The *Vibrio shiloi/Oculina patagonica* model system of coral bleaching. Ann Rev Microbiol 58:143–159

Shoguchi E et al (2012) Draft assembly of the Symbiodinium minutum nuclear genome reveals dinoflagellate gene structure. Curr Biol 23:1399–1408

Silverman J et al (2009) Coral reefs may start dissolving when atmospheric CO2 doubles. Geophys Res Lett 26:1–5

Silverstein RN et al (2012) Specificity is rarely absolute in coral-algal symbiosis: implications for coral response to climate change. Proc Royal Soc Ser B 279:2609–2618

Stolarski J et al (2011) The ancient evolutionary origins of Scleractinia revealed by azooxanthellate corals. BMC Evol Biol 11:316

Tchernov D et al (2004) Membrane lipids of symbiotic algae are diagnostic sensitivity to thermal bleaching in corals. Proc Natl Acad Sci U S A 101:13531–13535

Webster NS et al. (2013) Near-future ocean acidification causes differences in microbial associations within diverse coral reef taxa. Environmental Microbiology Reports 5:243–251

Ushijima B et al (2014) Vibrio coralliilyticus strain OCN008 is an etiological agent of acute *Montipora* white syndrome. Appl Environ Microbiol 80:2102–2109

The Hidden Impact of Viruses 10

Metaorganisms have evolved for more than 500 million years, yet only recently have advances in sequencing technology allowed us to appreciate the full nature of the complexities in host–microbe interactions. These technological advances have revealed that in addition to bacteria and associated eukaryotic symbionts, one more important player is an integral member of the holobiont: viruses (Fig. 10.1). Viruses are found anywhere cellular life exists, in any environment or ecosystem. They exist

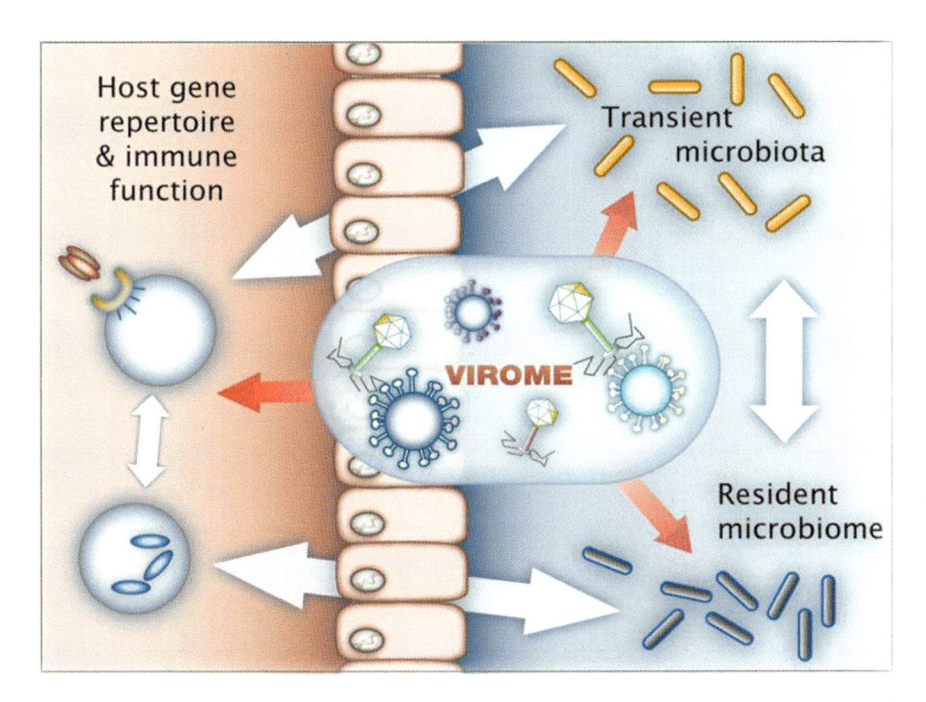

Fig. 10.1 Viruses and phages are likely to substantially impact shape the genetic diversity and the species diversity of the microbiota in the animal host

along with their hosts as part of a dynamic community ensemble of exogenous viral particles and endogenous proviruses. Yet, although being the most abundant and diverse biological component on the planet, comparatively little is known about their role in the holobiont. Does the virome play a role in establishing and maintaining the microbiome? What are the underlying mechanisms as to how viruses control microbes? Can bacteriophage-based therapies be used in invertebrates such as corals for treating complex disorders?

10.1 Beneficial Viruses

Metagenomic approaches have transformed our understanding of viruses in many ecosystems, significantly advancing our knowledge of diversity, abundance, and virus–host interactions. Outside of pathogenic effects, little is still known about viruses infecting eukaryotic hosts. In spite of the common perception of viruses as pathogens, many viruses are in fact beneficial to their hosts in various ways. Beneficial viruses have been discovered in many different hosts, including bacteria, insects, plants, fungi, and animals. For example, many parasitoid wasp species are known to harbor symbiotic viruses, and these viruses have a mutual relationship with their wasp host, particularly for host immune responses. The beneficial effects of viruses range from obligate mutualisms, in which the survival of the host is dependent on the virus, to benefits that occur only under specific environmental conditions. In addition, some of these relationships are ancient and the line between the virus and its host is blurry, and as the relationship between the aforementioned wasps and polydnaviruses has shown, some relationships are clearly symbiogenic. It seems that beneficial viruses have played a major part in the evolution of life on earth. However, how these beneficial interactions evolve is still a mystery in many cases.

A symbiotic function for viruses has only recently been demonstrated in mammals. Norovirus has been shown to compensate for the presence of the microbiome in germfree mice. A functional immune system was necessary for this compensation indicating that viral–host interactions are vital for animal health. Furthermore, the presence of bacteriophages in the mucus, in addition to host-secreted antimicrobial peptides, may allow for selection of the associated bacteria in the holobiont.

Phages likely have a profound impact on the composition and functional properties of the bacterial microbiota, which in turn could shape development and function of the immune system. Phages may serve as important reservoirs of genetic diversity in the microbiota by acting as vehicles for the horizontal transfer of virulence, antibiotic resistance, and metabolic determinants among bacteria. Bacterial acquisition of phage genes could modify the functional properties of the microbiota, thereby substantially impacting host metabolism and immunity. Phages may also impact the composition of commensal bacterial populations through bacteriophage-induced cell lysis, also termed "phage predation." Breck Duerkop and Lora Hooper from the University of Texas Southwestern Medical Center in Dallas have proposed that phage predation in fact may be widespread. The reason for this assumption is the presence of clustered regularly interspaced short palindromic repeat (CRISPR) systems in many human commensal bacteria (Duerkop and Hooper 2013). CRISPRs are genetic modules in which short repeats of foreign DNA are inserted between spacer sequences in the bacterial chromosome. The spacer sequences are transcribed into short complementary RNAs that target invading DNA for destruction. Since bacteria of the gastrointestinal tract and oral mucosa have acquired CRISPR-specific spacer sequences homologous to phage DNA, commensal bacteria most likely are preyed upon by lytic phages in the mammalian host.

10.2 Viral Communities in *Hydra* Are Species Specific and Sensitive to Stress

Using a metagenomic approach, researchers from San Diego State University in cooperation with a team of scientists at Kiel University in Germany recently have analyzed five different *Hydra* species, under non-stressed and heat-stressed conditions, for a total of ten viromes (Grasis et al. 2014). Taxonomic evaluation of the viral families isolated from lab-cultured or wild-caught *Hydra* revealed that each species of *Hydra* is associated with a diverse community of prokaryotic and eukaryotic viruses. Each of the *Hydra* species is characterized by associating with very distinct viral populations suggesting a phylosymbiotic pattern. The most common viral families associating with *Hydra* are prokaryotic viruses belonging to the *Myoviridae*, *Siphoviridae*, and *Inoviridae* families, as well as the eukaryotic viruses belonging to the *Herpesviridae* family. Outside of these families, a wide diversity of viruses unique to each species of *Hydra* was observed.

The largest percentages of associating viruses for each species of *Hydra* were found to be bacteriophages, which infect the bacterial and archaeal domains. Predicted prokaryotic hosts accounted for 38–63 % of the viral sequencing hits. The two most common families of bacteriophages associating with most *Hydra* species are the *Myoviridae* and *Siphoviridae* families. Similar results were achieved when looking into the viral communities found in the human gut, with large percentages of *Myoviridae* and *Siphoviridae* families associated with human individuals. These results indicate that the viruses associating with the host *Hydra* are not merely a reflection of the viruses present in the local water environment but that *Hydra* select specific viral communities. *Hydra*-associated viruses are not only species specific but also sensitive to temperature stress. In all investigated *Hydra* species, heat stress increased the diversity of viral families associated with the animal. Interestingly, in each species, heat stress also caused the diversity of bacteriophage hosts to increase. That temperature stress is causing increased host-associated viral diversity points to the presence of a diverse latent viral community, which can be reactivated by environmental stress to start the lytic part of the viral life cycle. It seems therefore that *Hydra* can dynamically regulate their associated viral communities to compensate for changing environmental factors. Indeed, functional metabolic profiles of the viromes associating with each species of *Hydra* show that many cellular functions in *Hydra* are affected by the increase in viral diversity upon heat stress (Fig. 10.2).

Fig. 10.2 Schematic view of the interactions between host organisms, their microbiota, and viruses (Taken from Bosch et al 2015)

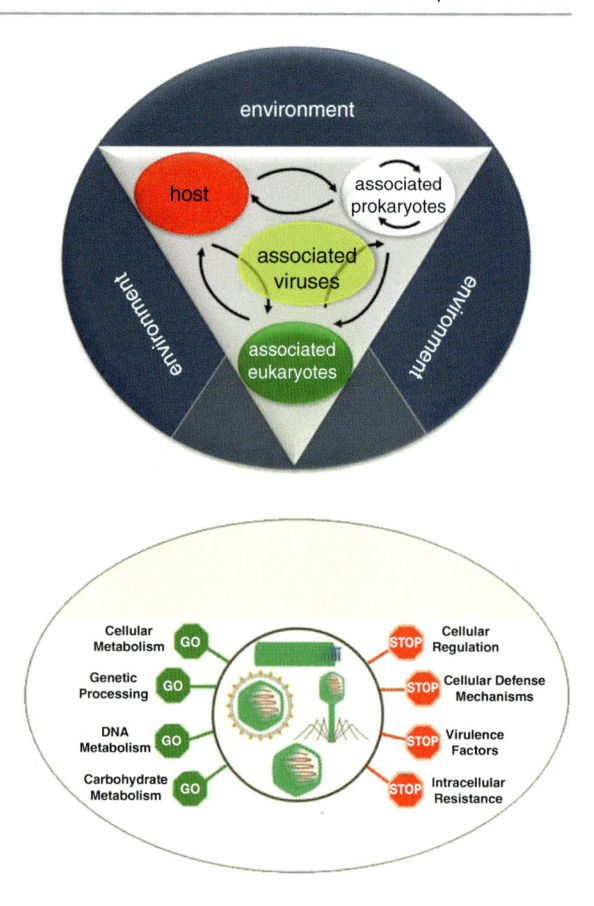

Specifically, cellular DNA and carbohydrate metabolism subsystems were increased with heat stress, while cellular virulence and defense mechanisms were nearly ablated. The increase of cellular metabolism and cellular genetic processing, as well as compromised cellular defense mechanisms, indicates viral involvement in the regulation of the host-associated microbiota under these conditions.

Aquatic environmental organisms like *Hydra* are permanently exposed to diverse bacteria and viruses from the surrounding plankton community. In contrast to terrestrial organisms, transmission of potential pathogens is independent from vector organisms and imposes an increased risk of viral host switching. Viral host switching is commonly observed in bacteriophages infecting individual hosts from different species and even general. In bacterial communities, phage predation has a tremendous impact on bacteria affecting 20–40 % of mortality. It is all the more astonishing that *Hydra* is able to preserve its specific bacterial and viral communities. The list of mechanisms involved include Hydra's innate immune system (see Chaps. 5 and 7), which is essential to establish and maintain a host-specific bacterial community. In addition, *Hydra*-specific bacteria may feature their own defense mechanisms to protect themselves from bacteriophage infection. Some *Hydra*-associated bacteria exhibit both CRISPR (*c*lustered *r*egularly *i*nterspaced *s*hort *p*alindromic *r*epeat) adaptive immunity (see Sect. 2.5) and restriction–modification (RM) system to degrade foreign DNA. Moreover, genome sequencing of *Hydra*-associated bacteria showed that prophage sequences exist in the genomes of the most abundant *Hydra*-specific bacteria. This is a first and truly exciting hint that phages may play a role in bacteria–bacteria interaction within the *Hydra* host. By exploiting the accessibility of the *Hydra* model system and assessing the impact of phages on bacterial growth of *Hydra*-specific bacteria as well as assessing the impact of phage–bacteria interactions in host colonization, we currently test the hypothesis that the presence of prophages prevents other bacteriophages from infecting and thus excludes superinfection and that prophages are important internal regulators to control the abundance of *Hydra*-associated bacteria.

Up to now, viruses were mostly considered as pathogenic agents. However, ideas about the role of viruses in multicellular organisms are changing and might be undergoing a paradigm shift. What is getting very clear is that environmental stress dramatically changes the interaction between host and viruses and that *Hydra* dynamically regulates this interaction and selectively shapes its virome to regain homeostasis. This suggests that genetic factors of the host can outweigh high mutation rates of viruses and select for "beneficial" viral variants. It seems, therefore, that we have to add viral replication and the capability of specific viromes for selection of both *Hydra* and associated bacteria hosts to the list of mechanisms used by early emerging holobionts to adapt to changing environmental conditions.

10.3   Bacteriophage Therapy in Corals?

Bacteriophage therapy aims to restore the normal symbiosis through specific lysis of the pathogenic bacteria, leading to increased homeostasis. Classically, the treatment uses a bacteriophage, or cocktail of several bacteriophages, to specifically lyse target pathogenic bacteria (Chanishvili 2012). As we have learned above, viruses and bacteriophages are also a common component of cnidarian holobionts. Intriguing studies from a team of scientists around Eugene Rosenberg at Tel Aviv University indicate that coral diseases such as coral white plague disease may be managed by bacteriophages to reduce bacterial infection (Efrony, et al. 2008). The rapidly progressing white plague-like disease of the coral *Favia favus* is caused by the gram-negative pathogen *Thalassomonas loyana*. In the control group, with no bacteriophage addition, the coral tissue infected with *T. loyana* was observed to be lysed after only 4–6 days. When corals were treated with bacteriophage BA3 1 day after *T. loyana* infection, they were protected from further infection for at least 37 days. However, corals inoculated with bacteriophage BA3 2 days after infection with *T. loyana* were not protected, indicating that bacteriophage therapy may be able to prevent the spread of the disease, rather than cure an already infected coral. A related study more recently investigated the feasibility of applying bacteriophage therapy to treat the coral pathogen *Vibrio coralliilyticus* (Cohen, et al. 2013). A specific bacteriophage for *V. coralliilyticus* strain P1, a lytic bacteriophage belonging to the *Myoviridae* family, was isolated from the seawater above corals. Bacteriophage therapy experiments using coral juveniles in microtiter plates as a model system revealed that this bacteriophage was able to prevent *V. coralliilyticus*-induced photoinactivation and tissue lysis.

Whether this kind of bacteriophage therapy becomes a mainstream approach to address the global problem of coral diseases remains to be shown. Major advantages of bacteriophage therapy of coral diseases include host specificity, self-replication, and environmental safety. The bacteriophage only attacks and destroys the specific pathogen, leaving the remaining beneficial microorganisms untouched. However, due to mutation rates of viruses, scientists stress that it is uncertain how bacteriophage therapy will progress in the field (Ben-Haim and Rosenberg 2002; Ben-Haim et al. 2003a, b; Cohen et al. 2013). They suggest "clinical field trials" to determine the efficacy of bacteriophage therapy in treatment of coral white plague disease. On the other hand, it is probably most important to understand the factors contributing to disease development and progression. In this context environmental stress has a fundamental effect on the stability and composition of the cnidarian holobiont, making it more susceptible to pathogen infections. Stress-induced changes, and the resulting imbalance of the holobiont, are important factors in elucidating complex diseases and not only in cnidarians.

Conclusion

By virtue of its morphological simplicity and molecular accessibility, *Hydra* allows thorough in vivo analysis of virus–bacteria–host interactions. Host factors required for virus colonization can be identified. The use of bacterial genetic tools and the isolation of *Hydra* bacteriophages from distinct bacterial hosts will help elucidate the specific microbe-derived factors.

References

Ben-Haim Y, Rosenberg E (2002) A novel Vibrio. sp. pathogen of the coral Pocillopora damicornis. Mar Biol 141:47–55

Ben-Haim Y, Zicherman-Keren M, Rosenberg E (2003a) Temperature-regulated bleaching and lysis of the coral Pocillopora damicornis by the novel pathogen Vibrio coralliilyticus. Appl Environ Microbiol 69(7):4236–4242

Ben-Haim Y, Thompson FL, Thompson CC, Cnockaert MC, Hoste B, Swings J, Rosenberg E (2003b) Vibrio coralliilyticus sp. nov., a temperature-dependent pathogen of the coral Pocillopora damicornis. Int J Syst Evol Microbiol 53(Pt 1):309–15

Bosch TC, Grasis JA, Lachnit T (2015) Microbial ecology in Hydra: why viruses matter. J Microbiol 53(3):193–200

Chanishvili N (2012) Phage therapy-history from Twort and d'Herelle through Soviet experience to current approaches. Adv Virus Res 83:3–40

Cohen Y, Pollock F, Rosenberg E, Bourne DG (2013) Phage therapy treatment of the coral pathogen Vibrio coralliilyticus. Microbiologyopen 2(1):64–74

Duerkop B, Hooper LV (2013) Resident viruses and their interactions with the immune system. Nat Immunol 14:654–659

Efrony R, Atad I, Rosenberg E (2008) Phage therapy of coral white plague disease: properties of phage BA3. Curr Microbiol 58:139–145

Grasis JA, Lachnit T, Anton-Erxleben F, Lim YW, Schmieder R, Fraune S, Franzenburg S, Insua S, Machado G, Haynes M, Little M, Kimble R, Rosenstiel P, Rohwer FL, Bosch TC (2014) Species-specific viromes in the ancestral holobiont Hydra. PLoS One 9:e109952

11. Seeking a Holistic View of Early Emerging Metazoans: The Power of Modularity

> *The time has come to replace the purely reductionist "eyes-down" molecular perspective with a new and genuinely holistic, eyes-up, view of the living world, one whose primary focus is on evolution, emergence, and biology's innate complexity*
>
> Carl R Woese (2004)

11.1 Animals Are Mobile Ecosystems Carrying a Myriad of Microbes with Them

In our tour of early emerging metazoans, we have seen that a coral colony may be rooted in a given place, but it is never lonely. There are symbiotic algae and bacteria living in, on, and near it. We also have seen that arranging a simple organism as a metaorganism of several self-coordinating parts can confer fitness and help to sustain periods of changing environmental conditions. We even came to the point to accept that life is fundamental multi-organismal. And we have seen some of the ways holobionts become fragile and break.

In our understanding, when first animals appeared very early in the history of life on earth (Chaps. 2 and 3), they were multicellular ecosystems which intimately coevolved with their microbes, such that even closely related animal species maintain unique microbiomes. These interactions occur in the context of conserved signaling pathways and mechanisms of tissue homeostasis (Chap. 5) whose detailed molecular logic remain elusive. The forces that shaped and still are shaping the colonizing microbial composition are the focus of much current investigation, and it is evident that there are pressures exerted both by the host and by the external environment to mold these ecosystems. Understanding the diversity of such genome–microbiome–environment interactions requires integrative, multidisciplinary, and modeling-based approaches (Bosch 2014; Gilbert et al. 2015).

Animals, therefore, are not considered individuals anymore by the traditional anatomical, physiological, immunological, genetic, or developmental accounts. Rather, evolution generates holobionts, organisms composed of numerous genetic lineages whose interactions are critical for the development and maintenance of the entire organism. The outcome of such relationships is sometimes observed as an evolutionary signal of host phylogeny and microbial community co-structuring between species – phylosymbiosis (Chap. 4).

The immune system seems to have evolved as a form of ecosystem management that exerts critical control over microbiota composition, diversity, and location (Chap. 5). We have demonstrated the critical importance of antimicrobial peptides and other components of the innate immune system in selecting a specific microbiome. Microbial communities may also have promoted the need for adaptive immunity in vertebrates. Therefore, both innate and acquired immunity appear to be involved in maintaining the animal holobiont.

Symbionts not only maintain species-specific interactions with their animal hosts but also may have promoted major evolutionary transitions and even can generate the conditions for reproductive isolation (Chap. 6). Across model systems, investigating microbial symbionts turns into a complex and multifactorial endeavor that requires carefully constructed, hypothesis-driven experimentation. The preceding chapters (in particular Chaps. 7 and 8) have shown that representative cnidarians and sponges provide novel opportunities to investigate how the environment and in particular how microbes interact with developmental processes and, in turn, how developmental evolution affects the environment.

Like the components of the immune system we discussed at the beginning of this book, the regulatory machinery controlling cell proliferation, differentiation, and tissue homeostasis has to respond to environmental changes and microbial exposure (Chap. 7). Stem cells don't live in sterile petri dishes but are exposed to heterogeneous environments. They communicate with their abiotic and biotic environment using stem cell-specific regulatory factors such as the transcription factor Foxo which in *Hydra* not only controls stem cells but also the functionality of the innate immune system. It becomes more and more apparent that studies of the development of animals cannot ignore the microbial world within. In fact it may be permitted to say that the study of development has been hampered by ignoring the multi-organismal nature of life. Symbiotic microbes and viruses are fundamental to nearly all aspects of host function and fitness (Gilbert et al. 2015).

Historically, we have not had the capacity to look at this as a system. But advances in genetic sequencing have opened up ways to survey entire holobiont communities. High-throughput transcriptome analysis, methylome analysis, polymerase chain reactions, and inexpensive DNA sequencing have allowed us in early emerging metazoans to discover a world of relationships that had not been seen before. Meanwhile, engineers and computational biologists have developed ways to manage large data sets and merge disparate recordings into cohesive models. What is needed now is to integrate not just the data about the host and the host's environment but all the biological components in that system. These new developments have given us an awareness of the dependency of phenotypes on environmental conditions as well as methods of studying them. The results have been totally unexpected and have changed our ideas about the "nature" of the simple organisms we are studying (Gilbert et al. 2015).

11.2 The Power of Modularity

Individuals, populations, and species all have to cope with environmental change by adapting physiologically through responses that are immediate and reversible. How? It is well known that organisms may respond by phenotypic plasticity, genetic adaptation, movement (range shifts), or extinction. Less explored is the impact of changes in biotic interactions and community structure on species' survival. We have seen in preceding chapters that *Hydra*'s associated microbiota including the associated viruses might change in response to environmental conditions (Chap. 7). The *Hydra* holobiont, therefore, appears to be a dynamic system being characterized by functional redundancy and fast adaptations to altered environmental conditions. It is this modularity and interoperability of the components of the holobiont which allows rapid adaptation to changing environmental conditions by altering the associated microbiota. This view is conceptualized by Rosenberg's "holobiont concept." Depending on the variety of different niches provided by the host, which can change with developmental stage, diet, or other environmental factors, a more or less diverse microbial community can be established within a given host species. The dynamic relationship between symbiotic microorganisms and environmental conditions results in the selection of the most advantageous holobiont.

It may be this kind of modular structure that provides corals with resistance against certain pathogens enabling them to adapt much faster to novel environmental conditions than by mutation and selection. Accordingly, we see host–microbe interactions as significant drivers of animal evolution and diversification (Chap. 6). As outlined in Chap. 7, the complexity of the holobiont is characterized by the fact that beneficial microbes represent a major factor whose activities always are tightly linked to tissue homeostasis, illustrated as stem cell factors, immunity, and nutrition (Fig. 11.1).

The strategies implied in these interactions are, at their core, strategies of resilience. Defining resilience is complicated by the fact that different fields use the term to mean slightly different things. We frame resilience in terms borrowed from both ecology and sociology as the capacity of a system to maintain its core purpose and integrity in the face of dramatically changed circumstances. The concept of resilience is a powerful lens through which we can view the multi-organismal nature of organisms afresh. Living organisms are messy and complex and robust but fragile at the same time; they are resilient systems rooted in diversity and difference and are tolerant of occasional dissent.

In the preceding pages, we have seen how arranging a living system as a holobiont and a network of self-coordinating parts can bolster its resilience. The newness of all of this microbiome research and the implications of these discoveries for revolutionizing many aspects of biology and medicine are truly exciting. And the lessons are clear: over the decades we have learnt about the toothed wheels, but we still do not understand the clock. The time has come for a holistic understanding of complex life processes.

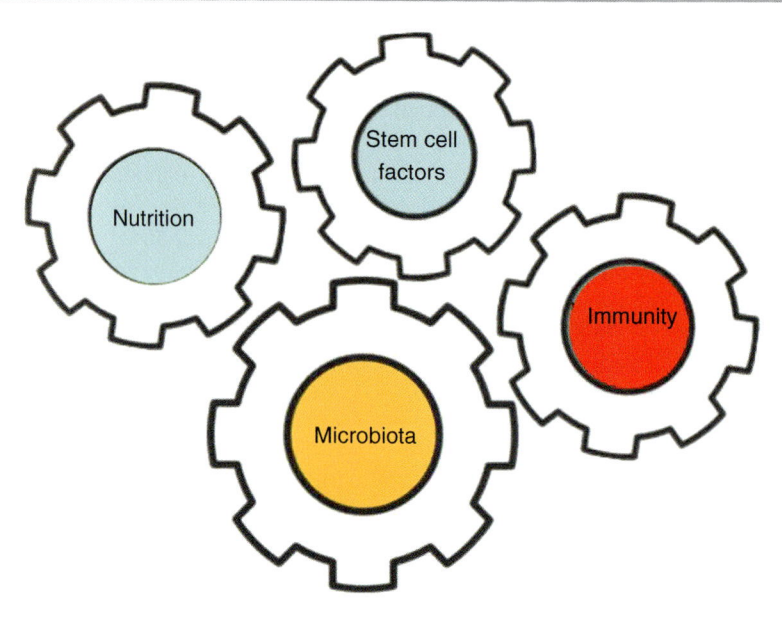

Fig. 11.1 In a holobiont, beneficial microbes represent a major factor whose activities are tightly linked to tissue homeostasis, illustrated as stem cell factors, immunity, and nutrition

References

Bosch TC (2014) Rethinking the role of immunity: lessons from Hydra. Trends Immunol 35(2014):495–502

Gilbert SF, Bosch TC, Ledón-Rettig C (2015) Eco-Evo-Devo: developmental symbiosis and developmental plasticity as evolutionary agents. Nat Rev Genet 16(10):611–622

Further Reading

Adamska M, Larroux C, Adamski M, Green K, Lovas E, Koop D, Richards GS, Zwafink C, Degnan BM (2010) Structure and expression of conserved Wnt pathway components in the demosponge Amphimedon queenslandica. Evol Dev 12(5):494–518

Adamska M, Degnan BM, Green K, Zwafink C (2011) What sponges can tell us about the evolution of developmental processes. Zoology (Jena) 114(1):1–10, Review

Ainsworth TD et al (2015) The coral core microbiome identifies rare bacterial taxa as ubiquitous endosymbionts. ISME J. doi:10.1038/ismej.2015.39

Alegado RA, King N (2014) Bacterial influences on animal origins. Cold Spring Harb Perspect Biol 6:a016162

Alegado RA, Brown LW, Cao S, Dermenjian RK, Zuzow R, Fairclough SR, Clardy J, King N (2012) A bacterial sulfonolipid triggers multicellular development in the closest living relatives of animals. Elife 1:e00013

Altura MA, Heath-Heckman EA, Gillette A, Kremer N, Krachler AM et al (2013) The first engagement of partners in the Euprymna scolopes–Vibrio fischeri symbiosis is a two-step process initiated by a few environmental symbiont cells. Environ Microbiol 15:2937–2950

Anctil M (2009) Chemical transmission in the sea anemone Nematostella vectensis: a genomic perspective. Comp Biochem Physiol D 4:268–289

Ardeshir A, Narayan NR, Méndez-Lagares G, Lu D, Rauch M, Huang Y, Van Rompay KK, Lynch SV, Hartigan-O'Connor DJ (2014) Breast-fed and bottle-fed infant rhesus macaques develop distinct gut microbiotas and immune systems. Sci Transl Med 6(252):252ra120

Artamonova II, Mushegiand AR (2013) Genome sequence analysis indicates that the model eukaryote Nematostella vectensis harbors bacterial consorts. Appl Environ Microbiol 79(22):6868–6873

Augustin R and Bosch TCG (2010) Cnidarian immunity: a tale of two barriers. In: Söderhäll K (ed) Invertebrate immunity landes bioscience and Springer Science+Business Media, LLC, New York, USA

Augustin R, Bosch TCG (2011) Cnidarian immunity: a tale of two barriers. Adv Exp Med Biol 708:1–16

Augustin R, Siebert S, Bosch TCG (2009a) Identification of a kazal-type serine protease inhibitor with potent anti-staphylococcal activity as part of Hydra's innate immune system. Dev Comp Immunol 33:830–837

Augustin R, Anton-Erxleben F, Jungnickel S, Hemmrich G, Spudy B, Podschun R, Bosch TCG (2009b) Activity of the novel peptide arminin against multiresistant human pathogens shows the considerable potential of phylogenetically ancient organisms as drug sources. Antimicrob Agents Chemother 53:5245–5250

Augustin R, Fraune S, Bosch TCG (2010) How Hydra senses and destroys microbes. Semin Immunol 22:54–58

Augustin R, Fraune S, Franzenburg S, Bosch TCG (2012) Where simplicity meets complexity: hydra, a model for host-microbe interactions. Adv Exp Med Biol 710:71–81

Bach JF (2002) The effect of infections on susceptibility to autoimmune and allergic diseases'. N Engl J Med 347(2002):911–920

Ball E, Hayward D, Saint R et al (2004) A simple plan – Cnidarians and the origins of developmental mechanisms. Nat Genet 5:567–577

Ball E, DeJong D, Schierwater B et al (2007) Implications of cnidarian gene expression patterns for the origins of bilaterality – is the glass half full or half empty? Integr Comp Biol 47(5):701–711

Barash Y, Sulam R, Loya Y, Rosenberg E (2005) Bacterial Strain BA-3 and a filterable factor cause a white plague-like disease in corals from the Eilat coral reef. Aquat Microbiol Ecol 40:183–189

Barr JJ, Auro R, Furlan M, Whiteson KL, Erb ML, Pogliano J, Stotland A, Wolkowicz R, Cutting AS, Doran KS, Salamon P, Youle M, Rohwer F (2013) Bacteriophage adhering to mucus provide a non-host-derived immunity. Proc Natl Acad Sci U S A 110(26):10771–10776

Barrangou R, Fremaux C, Deveau H, Richards M, Boyaval P, Moineau S, Romero DA, Horvath P (2007) CRISPR provides acquired resistance against viruses in prokaryotes. Science 315:1709–1712

Barton ES, White DW, Cathelyn JS, Brett-McClellan KA, Engle M, Diamond MS, Virgin HW (2007) Herpesvirus latency confers symbiotic protection from bacterial infection. Nature 447:326–329

Bates JM, Mittge E, Kuhlman J, Baden KN, Cheesman SE, Guillemin K (2006) Distinct signals from the microbiota promote different aspects of zebrafish gut differentiation. Dev Biol 297(2):374–386

Bell G (1998) Model metaorganism. A book review. Science 282(5387):248

Ben-Haim Y, Rosenberg E (2002) A novel Vibrio sp. pathogen of the coral Pocillopora damicornis. Mar Biol 141:47–55

Ben-Haim Y, Zicherman-Keren M, Rosenberg E (2003) Temperature-regulated bleaching and lysis of the coral Pocillopora damicornis by the novel pathogen Vibrio coralliilyticus. Appl Environ Microbiol 69(7):4236–4242

Ben-Haim Y, Thompson FL, Thompson CC, Cnockaert MC, Hoste B, Swings J, Rosenberg E (2013) Vibrio coralliilyticus sp. nov., a temperature-dependent pathogen of the coral Pocillopora damicornis. Int J Syst Evol Microbiol 53(Pt 1):309–315

Bevins CL, Salzman NH (2011) The potter's wheel: the host's role in sculpting its microbiota. Cell Mol Life Sci 68:3675–3685

Biagi E, Candela M, Fairweather-Tait S, Franceschi C, Brigidi P (2012) Ageing of the human metaorganism: the microbial counterpart. Age (Dordr) 34(1):247–267

Blanc G et al (2010) The Chlorella variabilis NC64A genome reveals adaptation to photosymbiosis, coevolution with viruses, and cryptic sex. Plant Cell 22:2943–2955

Blaser MJ (2014) The microbiome revolution. J Clin Invest 124(10):4162–4165

Boehm AM, Hemmrich G, Khalturin K, Puchert M, Anton-Erxleben F, Wittlieb J, Klostermeier UC, Rosenstiel P, Oberg H-H, Bosch TCG (2012) FoxO is a critical regulator of stem cell maintenance and immortality in Hydra. Proc Natl Acad Sci U S A 109(48):19697–19702

Boehm AM, Rosenstiel P, Bosch TCG (2013) Stem cells and aging from a quasi-immortal point of view. BioEssays 35(11):994–1003

Bordenstein SR (2014) Genomic and cellular complexity from symbiotic simplicity. Cell 158(6):1236–1237

Bordenstein SR, Theis KR (2015) Host biology in light of the microbiome: ten principles of holobionts and hologenomes. PLoS Biol 13(8):e1002226

Bosch TCG (2007) Why polyps regenerate and we don't: towards a cellular and molecular framework for Hydra regeneration. Dev Biol 303:421–433

Bosch TCG (2012a) What Hydra has to say about the role and origin of symbiotic interactions. Biol Bull 223:78–84

Bosch TCG (2012b) Understanding complex host-microbe interactions in Hydra. Gut Microb 3(4):1–7

Bosch TCG (2013) Cnidarian-microbe interactions and the origin of innate immunity in metazoans. Ann Rev Microbiol 67:499–518, pdficon

Bosch TCG (2014) Rethinking the role of immunity: lessons from Hydra. Trends Immunol 35(2014):495–502
Bosch TCG, David CN (1986) Immunocompetence in hydra: epithelial cells recognize self-nonself and react against it. J Exp Biol 238:225–234
Bosch TCG, McFall-Ngai M (2011) Metaorganisms as the new frontier. Zoology 114(2011):185–190
Bosch TCG, Augustin R, Anton-Erxleben F, Fraune S, Hemmrich G, Zill H, Rosenstiel P, Jacobs G, Schreiber S, Leippe M, Stanisak M, Grotzinger J, Jung S, Podschun R, Bartels J, Harder J, Schroder JM (2009) Uncovering the evolutionary history of innate immunity: the simple metazoan Hydra uses epithelial cells for host defence. Dev Comp Immunol 33:559–569
Bosch TCG, Adamska M, Augustin R, Domazet-Loso T, Foret S, Fraune S, Funayama N, Grasis J, Hamada M, Hatta M, Hobmayer B, Kawai K, Klimovich A, Manuel M, Shinzato C, Technau U, Yum S, Miller DJ (2014) How do environmental factors influence life cycles and development? An experimental framework for early-diverging metazoans. Bioessays 36(12):1185–1194
Bosch TCG, Grasis JA, Lachnit T (2015) Microbial ecology in Hydra: why viruses matter. J Microbiol 53(3):193–200
Bourne D et al (2008) Changes in coral-associated microbial communities during a bleaching event. ISME J 2:350–363
Bourne DG et al (2009) Microbial disease and the coral holobiont. Trends Microbiol 17:554–562
Bourne DG et al (2013a) Coral reef invertebrate microbiomes correlate with the presence of photosymbionts. ISME J 7:1452–1458
Bourne D, Webster NS et al (2013b) Coral reef bacterial communities. In: Rosenberg E et al (eds) The prokaryotes – prokaryotic communities and ecophysiology. Springer, Berlin
Broadbent AD, Jones GB (2004) DMS and DMSP in mucus ropes, coral mucus, surface films and sediment pore waters from coral reefs of the Great Barrier Reef. Mar Freshw Res 55: 849–855
Brodie J et al (2005) Are increased nutrient inputs responsible for more outbreaks of crown of thorns starfish – an appraisal of the evidence. Mar Pollut Bull 51:266–278
Brucker RM, Bordenstein SR (2012a) Speciation by symbiosis. Trends Ecol Evol 27(8):443–451
Brucker RM, Bordenstein SR (2012b) The roles of host evolutionary relationships (genus: Nasonia) and development in structuring microbial communities. Evolution 66(2):349–362
Brucker RM, Bordenstein SR (2013a) The hologenomic basis of speciation: gut bacteria cause hybrid lethality in the genus Nasonia. Science 341(6146):667–669
Brucker RM, Bordenstein SR (2013b) The capacious hologenome. Zoology (Jena) 116(5):260–261
Brucker RM, Bordenstein SR (2014) Response to Comment on "The hologenomic basis of speciation: gut bacteria cause hybrid lethality in the genus Nasonia". Science 345:1011
Buddemeier RW, Fautin DG (1993) Coral bleaching as an adaptive mechanism. BioScience 43:320–326
Campbell RD (1990) Transmission of symbiotic algae through sexual reproduction in Hydra: movement of algae into the oocyte. Tissue Cell 22:137–147
Cartwright P, Collins A (2007) Fossils and phylogenies: integrating multiple lines of evidence to investigate the origin of early major metazoan lineages. Integr Comp Biol 47:744–751
Chanishvili N (2012) Phage therapy-history from Twort and d'Herelle through Soviet experience to current approaches. Adv Virus Res 83:3–40
Chapman JA, Kirkness EF, Simakov O, Hampson SE, Mitros T, Weinmaier T, Rattei T, Balasubramanian PG, Borman J, Busam D, Disbennett K, Pfannkoch C, Sumin N, Sutton GG, Viswanathan LD, Walenz B, Goodstein DM, Hellsten U, Kawashima T, Prochnik SE, Putnam NH, Shu S, Blumberg B, Dana CE, Gee L, Kibler DF, Law L, Lindgens D, Martinez DE, Peng J, Wigge PA, Bertulat B, Guder C, Nakamura Y, Ozbek S, Watanabe H, Khalturin K, Hemmrich G, Franke A, Augustin R, Fraune S, Hayakawa E, Hayakawa S, Hirose M, Hwang JS, Ikeo K, Nishimiya-Fujisawa C, Ogura A, Takahashi T, Steinmetz PR, Zhang X, Aufschnaiter R, Eder MK, Gorny AK, Salvenmoser W, Heimberg AM, Wheeler BM, Peterson KJ, Bottger A, Tischler P, Wolf A, Gojobori T, Remington KA, Strausberg RL, Venter JC, Technau U, Hobmayer B, Bosch TC, Holstein TW, Fujisawa T, Bode HR, David CN, Rokhsar DS, Steele RE (2010) The dynamic genome of Hydra. Nature 464(7288):592–596

Chaston J, Douglas AE (2012) Making the most of "omics" for symbiosis research. Biol Bull 223(1):21–29, pmid:22983030

Chen Z, Wang A (2008) A new species of the genus Hydra from China (Hydrozoa, Hydraridae). Acta Zootaxonomica Sin 33:737–741

Chen J-Y, Oliveri P, Gao F, Dornbos SQ, Li C-W, Bottjer DJ, Davidson EH (2002) Precambrian animal life: Probable developmental and adult cnidarian forms from southwest China. Dev Biol 248:182–196

Cho CE, Norman M (2013) Cesarean section and development of the immune system in the offspring. Am J Obstet Gynecol 208(4):249–254

Chu H, Mazmanian SK (2013) Innate immune recognition of the microbiota promotes host-microbial symbiosis. Nat Immunol 14(7):668–675

Clarke G, Grenham S, Scully P, Fitzgerald P, Moloney RD, Shanahan F, Dinan TG, Cryan JF (2013) The microbiome-gut–brain axis during early life regulates the hippocampal serotonergic system in a sex-dependent manner. Mol Psychiatry 18:666–673

Clarke G, Stilling RM, Kennedy PJ, Stanton C, Cryan JF, Dinan TG (2014) Minireview: Gut microbiota: the neglected endocrine organ. Mol Endocrinol 28(8):1221–1238

Coffroth MA, Santos SR (2005) Genetic diversity of symbiotic dinoflagellates in the genus Symbiodinium. Protist 156:19–34

Cohen Y, Pollock F, Rosenberg E, Bourne DG (2013) Phage therapy treatment of the coral pathogen Vibrio coralliilyticus. Microbiol Open 2(1):64–74

Colman RJ, Rubin DT (2014) Fecal microbiota transplantation as therapy for inflammatory bowel disease: a systematic review and meta-analysis. J Crohn's Colitis 8(12):1569–1581 (1)

Cook CB, Kelty MO (1982) Glycogen, protein, and lipid content of green, aposymbiotic and nonsymbiotic hydra during starvation. J Exp Zool 222:1–9

Cullen TW et al (2015) Gut microbiota. Antimicrobial peptide resilience of prominent gut commensals during inflammation. Science 347(6218):170–175

Currie CR (2001) A community of ants, fungi, and bacteria: a multilateral approach to studying symbiosis. Annu Rev Microbiol 55:357–380

Dasgupta S, Kasper DL (2013) Relevance of commensal microbiota in the treatment and prevention of inflammatory bowel disease. Inflamm Bowel Dis 19:2478–2489

Davies J (2001) In a map for human life, count the microbes, too. Science 291:2316

Davy SK et al (2012) Cell biology of cnidarian-dinoflagellate symbiosis. Microbiol Mol Biol Rev 76:229–261

Dayel MJ, Alegado RA, Fairclough SR, Levin TC, Nichols SA, McDonald K, King N (2011) Cell differentiation and morphogenesis in the colony-forming choanoflagellate Salpingoeca rosetta. Dev Biol 357(1):73–82

De'ath G et al (2012) The 27-year decline of coral cover on the Great Barrier Reef and its causes. Proc Natl Acad Sci U S A 109:17995–17999

Dedeine F, Vavre F, Fleury F, Loppin B, Hochberg ME, Boulétreau M (2001) Removing symbiotic Wolbachia specifically inhibits oogenesis in a parasitic wasp. Proc Natl Acad Sci USA 98:6247–6252

Desbonnet L, Clarke G, Shanahan F, Dinan TG, Cryan JF (2014) Microbiota is essential for social development in the mouse. Mol Psychiatry 19(2):146–148

Dominguez-Bello MG et al (2010) Delivery mode shapes the acquisition and structure of the initial microbiota across multiple body habitats in newborns. Proc Natl Acad Sci U S A 107:11971–11975

Doolittle WF, Zhaxybayeva O (2009) On the origin of prokaryotic species. Genome Res 19(5):744–756

Doolittle WF, Fraser P, Gerstein MB, Graveley BR, Henikoff S, Huttenhower C, Oshlack A, Ponting CP, Rinn JL, Schatz MC, Ule J, Weigel D, Weinstock GM (2013) Sixty years of genome biology. Genome Biol 14(4):113

Douglas AE (2003) Coral bleaching – how and why? Mar Pollut Bull 46:385–392

Douglas AE (2014) Symbiosis as a general principle in eukaryotic evolution. Cold Spring Harb Perspect Biol 6(2):13

Douglas AE (2015) Multiorganismal insects: diversity and function of resident microorganisms. Annu Rev Entomol 60(1):17–34

Du Z, Zhang W, Xia H, Lü G, Chen G (2010) Isolation and diversity analysis of heterotrophic bacteria associated with sea anemones. Acta Oceanol Sin 29:62–69

Dubilier N, Bergin C, Lott C (2008) Symbiotic diversity in marine animals: the art of harnessing chemosynthesis. Nat Rev Microbiol 6(10):725

Duerkop B, Hooper LV (2013) Resident viruses and their interactions with the immune system. Nat Immunol 14:654–659

Dunbar HE, Wilson ACC, Ferguson NR, Moran NA (2007) Aphid thermal tolerance is governed by a point mutation in bacterial symbionts. PLoS Biol 5:e96

Eberl G (2010) A new vision of immunity: homeostasis of the superorganism. Mucosal Immunol 3:450–460

Efrony R, Atad I, Rosenberg E (2008) Phage therapy of coral white plague disease: properties of phage BA3. Curr Microbiol 58:139–145

Ehrlich PR, Raven PH (1964) Butterflies and Plants: a Study in Coevolution. Evolution 18:586–608

Eisenhauer N, Schlaghamersky J, Reich PB, Frelich LE (2011) The wave towards a new steady state: effects of earthworm invasion on soil microbial functions. Biol Invasions 13:2191–2196

Eitel M et al (2013) Global diversity of the Placozoa. PLoS One 8:e57131

Eme L, Doolittle WF (2015) Microbial diversity: a bonanza of phyla. Curr Biol 25(6):R227–R230

Erwin DH, Valentine JW (2013) The Cambrian Explosion. The construction of animal biodiversity. Roberts and Company, Greenwood Village

Erwin DH, Laflamme M, Tweedt SM, Sperling EA, Pisani D, Peterson KJ (2011) The Cambrian conundrum: early divergence and later ecological success in the early history of animals. Science 334:1091–1097

Ezenwa VO, Gerardo NM, Inouye DW, Medina M, Xavier JB (2012) Animal behavior and the microbiome. Science 338:198–199

Fairclough SR, Chen Z, Kramer E, Zeng Q, Young S, Robertson HM, Begovic E, Richter DJ, Russ C, Westbrook MJ, Manning G, Lang BF, Haas B, Nusbaum C, King N (2013) Premetazoan genome evolution and the regulation of cell differentiation in the choanoflagellate Salpingoeca rosetta. Genome Biol 14:R15

Feuda R, Hamilton SC, McInerney JO, Pisani D (2012) Metazoan opsin evolution reveals a simple route to animal vision. Proc Natl Acad Sci U S A 109:18868–18872

Fortunato SAV, Adamski M, Mendivil Ramos O, Leininger S, Liu J, Ferrier DEK, Adamska M (2014) Calcisponges have a ParaHox gene and dynamic expression of dispersed NK homeobox genes. Nature 514:620–623

Franzenburg S, Walter J, Künzel S, Baines JF, Bosch TCG, Fraune S (2013a) Distinct antimicrobial tissue activity shapes host species-specific bacterial associations. Proc Natl Acad Sci U S A 110(39):E3730–E3738

Franzenburg S, Fraune S, Altrock PM, Künzel S, Baines JF, Traulsen A, Bosch TCG (2013b) Bacterial colonization of Hydra hatchlings follows a robust temporal pattern. ISME J 7(4):781–790

Fraune S, Bosch TCG (2007) Long-term maintenance of species-specific bacterial microbiota in the basal metazoan Hydra. Proc Natl Acad Sci U S A 104:13146–13151

Fraune S, Bosch TCG (2010) Why bacteria matter in animal development and evolution. Bioessays 32:571–580

Fraune S, Bosch TCG (2012) Host-symbiont interactions: why Cnidaria matter. Cell News 3(2012):14–21

Fraune S, Abe Y, Bosch TCG (2009a) Disturbing epithelial homeostasis in the metazoan Hydra leads to drastic changes in associated microbiota. Environ Microbiol 11(9):2361–2369

Fraune S, Augustin R, Bosch TCG (2009b) Exploring host-microbe interactions in hydra. Microbe 4(10):457–462

Fraune S, Augustin R, Anton-Erxleben F, Wittlieb J, Gelhaus C, Klimovich VB, Samoilovich MP, Bosch TCG (2010) In an early branching metazoan, bacterial colonization of the embryo is controlled by maternal antimicrobial peptides. Proc Natl Acad Sci USA 107:18067–18072

Fraune S, Augustin R, Bosch TCG (2011a) Embryo protection in contemporary immunology: why bacteria matter. Commun Integr 4:369–372

Fraune S, Franzenburg S, Augustin R, Bosch TCG (2011b) Das Prinzip Metaorganismus. Biospektrum 17(6):634–636

Fraune S, Anton-Erxleben F, Augustin R, Franzenburg S, Knop M, Schröder K, Willoweit-Ohl D, Bosch TCG (2015) Bacteria-bacteria interactions within the microbiota of the ancestral metazoan Hydra contribute to fungal resistance. ISME J 9:1543–1556. doi:10.1038/ismej.2014.239

Fukami T, Wardle DA, Bellingham PJ, Mulder CP, Towns DR, Yeates GW, Bonner KI, Durrett MS, Grant-Hoffman MN, Williamson WM (2006) Above- and below-ground impacts of introduced predators in seabird-dominated island ecosystems. Ecol Lett 9:1299–1307

Futuyma DJ (1979) Evolutionary biology, 1st edn. Sinauer Associates, Sunderland. ISBN 0-87893-199-6

Futuyma DJ (1998) Evolutionary biology, 3rd ed. Sinauer Associates, Sunderland. (dated 1998, published 1997) ISBN 0-87893-189-9

Futuyma DJ (2005) Evolution. Sinauer Associates, Sunderland. ISBN 0-87893-187-2

Futuyma DJ, Slatkin M (eds) (1983) Coevolution. Sinauer Associates, Sunderland. ISBN 0-87893-228-3

Garm A, Mori S (2009) Multiple photoreceptor systems control the swim pacemaker activity in box jellyfish. J Exp Biol 212:3951–3960

Garren M et al (2014) A bacterial pathogen uses dimethylsulfoniopropionate as a cue to target heat-stressed corals. ISME J 8:999–1007

Gilbert SF (2014) Symbiosis as the way of eukaryotic life: the dependent co-origination of the body. J Biosci 39(2):201–209

Gilbert SF, Epel D (2015) ecological developmental biology: the environmental regulation of development, health, and evolution. Sinauer Associates Inc., Sunderland

Gilbert SF, Sapp J, Tauber AI (2012) A symbiotic view of life: we have never been individuals. Q Rev Biol 87(4):325–341

Gilbert SF, Bosch TCG, Ledón-Rettig C (2015) Eco-Evo-Devo: developmental symbiosis and developmental plasticity as evolutionary agents. Nat Rev Genet 16(10):611–622

Gordon J, Knowlton N, Relman DA, Rohwer F, Youle M (2013a) Superorganisms and holobionts. Microbe 8(4):152–153

Gordon J, Knowlton N, Relman DA, Rohwer F, Youle M (2013) Superorganisms and holobionts. Microbe Magazine, April 2013 issue

Grasis JA, Lachnit T, Anton-Erxleben F, Lim YW, Schmieder R, Fraune S, Franzenburg S, Insua S, Machado G, Haynes M, Little M, Kimble R, Rosenstiel P, Rohwer FL, Bosch TCG (2014) Species-specific viromes in the ancestral holobiont Hydra. PLoS One 9:e109952

Grissa I, Vergnaud G, Pourcel C (2007) The CRISPRdb database and tools to display CRISPRs and to generate dictionaries of spacers and repeats. BMC Bioinform 23(8):172

Grmek MD (1969) Préliminaires d'une étude historique des maladies. Annales Economie Société Civilisation 24:1437–1483

Guppy R, Bythell JC (2006) Environmental effects on bacterial diversity in the surface mucus layer of the reef coral *Montastraea faveolata*. Mar Ecol Prog Ser 328:133–142

Haapkyla J et al (2011) Seasonal rainfall and runoff promote coral disease on an inshore reef. PLoS One 6:e16893

Haber M (2013) Colonies are individuals: revisiting the superorganism revival. In: Bouchard F, Huneman P (eds) From groups to individuals: evolution and emerging individuality. The MIT Press, Cambridge, pp 195–217

Habetha M, Bosch TCG (2005) Symbiotic Hydra express a plant-like peroxidase gene during oogenesis. J Exp Biol 208:2157–2164

Habetha M, Anton-Erxleben F, Neumann K, Bosch TCG (2003) The Hydra viridis/Chlorella symbiosis. (I) Growth and sexual differentiation in polyps without symbionts. Zoology 106(2):101–108

Hadfield MG (2011) Biofilms and marine invertebrate larvae: what bacteria produce that larvae use to choose settlement sites. Ann Rev Mar Sci 3:453–470

Weismann to Haeckel, 27 January 1874, in Georg Uschmann and Bernhard Hassenstein (1965) Der Briefwechsel zwischen Ernst Haeckel und August Weismann. In: Gersch M (ed) Kleine Festrede aus Anlass der hundertjährigen Wiederklehr der gründung des Zoologischen Institues der Friedrich-Schiller-Universität Jena im Jahre 1865. Friedrich-Schiller-Universität, Jena, pp 35–36

Hamann O (1882) Zur Entstehung und Entwicklung der grünen Zellen bei Hydra. Z Wiss Zool 37:457–464

Hemmrich G, Bosch TCG (2008) Compagen, a comparative genomics platform for early branching metazoan animals, reveals early origins of genes regulating stem cell differentiation. Bioessays 20(10):1010–1018

Hemmrich G, Miller DJ, Bosch TCG (2007) The evolution of immunity – a low life perspective. Trends Immunol 28(10):449–454

Hemmrich G, Khalturin K, Boehm AM, Puchert M, Anton-Erxleben F, Wittlieb J, Klostermeier UC, Rosenstiel P, Oberg HH, Domazet-Lošo T, Sugimoto T, Niwa H, Bosch TCG (2012) Molecular signatures of the three stem cell lineages in Hydra and the emergence of stem cell function at the base of multicellularity. Mol Biol Evol. doi:10.1093/molbev/mss13

Hoerauf A (2008) Mansonella perstans—the importance of an endosymbiont. N Engl J Med 361:1502–1504

Hooper LV, Gordon JI (2001) Commensal host-bacterial relationships in the gut. Science 292(5519):1115–1118

Horvath P, Barrangou R (2010) CRISPR/Cas, the immune system of bacteria and archaea. Science 327:167–170

Hsiao EY, McBride SW, Hsien S, Sharon G, Hyde ER, McCue T, Codelli JA, Chow J, Reisman SE, Petrosino JF, Patterson PH, Mazmanian SK (2013) Microbiota modulate behavioral and physiological abnormalities associated with neurodevelopmental disorders. Cell 155(7):1451–1463

Jager M, Dayraud C, Mialot A, Quéinnec E, le Guyader H et al (2013) Evidence for involvement of Wnt signalling in body polarities, cell proliferation, and the neuro-sensory system in an adult ctenophore. PLoS One 8(12):e84363.

Jolley E, Smith DC (1978) The green hydra symbiosis. I. Isolation culture and characteristics of the chlorella symbiont of the "european" hydra viridis 1978. New Phytol 81:637–645

Jung S, Dingley AJ, Augustin R, Anton-Erxleben F, Stanisak M, Gelhaus C, Gutsmann T, Hammer MU, Podschun R, Bonvin AM, Leippe M, Bosch TCG, Grotzinger J (2009) Hydramacin-1, structure and antibacterial activity of a protein from the basal metazoan Hydra. J Biol Chem 284:1896–1905

Kaufman LS (1983) Effects of Hurricane Allen on reef fish assemblages near Discovery Bay, Jamaica. Coral Reefs 2:1–5

Kawaida H, Ohba K, Koutake Y, Shimizu H, Tachida H, Kobayakawa Y (2013) Symbiosis between Hydra and Chlorella: molecular phylogenetic analysis and experimental study provide insight into its origin and evolution. Mol Phylogenet Evol 66:906–914

Keeling PJ, McCutcheon JP, Doolittle WF (2015) Symbiosis becoming permanent: survival of the luckiest. Proc Natl Acad Sci U S A 112(33):10101–10103

Kernbauer E, Ding Y, Cadwell K (2014) An enteric virus can replace the beneficial function of commensal bacteria. Nature. doi:10.1038/nature13960

Khalturin K, Anton-Erxleben F, Sassmann S, Wittlieb J, Hemmrich G, Bosch TCG (2008) A novel gene family controls species-specific morphological traits in Hydra. PLoS Biol 6:e278

Khalturin K, Hemmrich G, Fraune S, Augustin R, Bosch TCG (2009) More than just orphans: are taxonomically-restricted genes important in evolution? Trends Genet 25:404–413

Khosravi A, Yáñez A, Price JG, Chow A, Merad M, Goodridge HS, Mazmanian SK (2014) Gut microbiota promote hematopoiesis to control bacterial infection. Cell Host Microbe 15(3):374–381

King N (2004) The unicellular ancestry of animal development. Dev Cell 7:313–325

King N (2005) Choanoflagellates. Curr Biol 15(4):R113–R114

King N (2010) Nature and nurture in the evolution of cell biology. Mol Biol Cell 21:3801–3802

King N, Young SL, Abedin M, Carr M, Leadbeater BSC (2008) The choanoflagellates: heterotrophic nanoflagellates and the sister group of the Metazoa in Emerging Model Organisms, vol. 1. Cold Spring Harbor Laboratory Press, Cold Spring Harbor, NY

King N, Westbrook M, Young SL, Kuo A, Abedin M, Chapman J, Fairclough S, Hellsten U, Isogai Y, Letunic I, Marr M, Pincus D, Putnam N, Rokas A, Wright KJ, Zuzow R, Dirks W, Good M, Goodstein D, Lemons D, Li W, Lyons J, Morris A, Nichols S, Richter DJ, Salamov A, Sequencing JGI, Bork P, Lim WA, Manning G, Miller WT, McGinnis W, Shapiro H, Tjian R, Grigoriev IV, Rokhsar D (2008b) The genome of the choanoflagellate Monosiga brevicollis and the origin of metazoans. Nature 451:783–788

Klueter A et al (2015) Taxonomic and environmental variation of metabolite profiles in marine dinoflagellates of the genus *Symbiodinium*. Metabolites 5:74–99

Knoll AH (2003) Life on a young planet. Princeton University Press, Princeton

Knowlton N (2001) The future of coral reefs. Proc Natl Acad Sci U S A 98(10):5419–5425

Knowlton N, Rohwer F (2003) Multispecies microbial mutualisms on coral reefs: the host as a habitat. Am Nat 162:S51–S62

Kortschak RD et al (2003) EST analysis of the cnidarian Acropora millepora reveals extensive gene loss and rapid sequence divergence in the model invertebrates. Curr Biol 13(24):2190–2195

Kozmik Z et al (2008) Assembly of the cnidarian camera-type eye from vertebrate-like components. Proc Natl Acad Sci U S A 105:8989–8993

Kremer N, Philipp EE, Carpentier MC, Brennan CA, Kraemer L et al (2013) Initial symbiont contact orchestrates host-organ-wide transcriptional changes that prime tissue colonization. Cell Host Microbe 14:183–194

Kushmaro A, Loya Y, Fine M, Rosenberg E (1996) Bacterial infection and coral bleaching. Nature 380:396–403

Kushmaro A, Rosenberg E, Fine M, Loya Y (1997) Bleaching of the coral Oculina patagonica by Vibrio AK-1. Mar Ecol Prog Ser 147:159–165

Kushmaro A, Rosenberg E, Fine M, Ben-Haim Y, Loya Y (1998) Effect of temperature on bleaching of the coral Oculina patagonica by Vibrio shiloi AK-1. Marine Ecol Prog Ser 171:131–137

Kvennnefors ECF, Roff G (2009) Evidence of cyanobacteria-like endosymbioints in Acroporid corals from the Great Barrier Reef. Coral Reefs 28:547

Landmann F, Foster JM, Michalski ML, Slatko BE, Sullivan W (2014) Co-evolution between an endosymbiont and its nematode host: Wolbachia asymmetric posterior localization and AP polarity establishment. PLoS Negl Trop Dis 8(8):e3096

Lange C, Hemmrich G, Klostermeier UC, López-Quintero JA, Miller DJ, Rahn T, Weiss Y, Bosch TCG, Rosenstiel P (2011) Defining the origins of the NOD-like receptor system at the base of animal evolution. Mol Biol Evol 28:1687–1702

Lederberg J (1996) Smaller fleas … ad infinitum: therapeutic bacteriophage redux. Proc Natl Acad Sci U S A 93:3167–3168

Leininger S, Adamski M, Bergum B, Guder C, Liu J, Laplante M, Bråte J, Hoffmann F, Fortunato S, Jordal S, Rapp HT, Adamska M (2014) Developmental gene expression provides clues to relationships between sponge and eumetazoan body plans. Nat Comm. doi:10.1038/ncomms4905

Lema KA et al (2012) Corals form characteristic associations with symbiotic nitrogen-fixing bacteria. Appl Environ Microbiol 78:3136–3144

Lema KA et al (2014) Onset and establishment of diazotrophs and other bacterial associates in the early life-history stages of the coral *Acropora millepora*. Mol Ecol 23:4682–4695

Lenhoff HM, Muscatine L (1963) On the role of algae symbiotic with Hydra. Science 142:956–958

Lesser MP et al (2004) Discovery of symbiotic nitrogen-fixing cyanobacteria in corals. Science 305:997–1000

Lesser MP et al (2007) Nitrogen fixation by symbiotic cyanobacteria provides a source of nitrogen for the scleractinian coral Montastraea cavernosa. Mar Ecol Prog Ser 346:143–152

Levy O et al (2011) Complex diel cycles of gene expression in coral-algal symbiosis. Science 331:175

Li X, Feng J, Sun R (2011) Oxidative stress induces reactivation of Kaposi's sarcoma-associated herpesvirus and death of primary effusion lymphoma cells. J Virol 85:715–724

Li X-Y, Pietschke C, Fraune S, Altrock PM, Bosch TCG, Traulsen A (2015) Which games are growing bacterial populations playing? J R Soc Interface. doi:10.1098/rsif.2015.0121

Littman RA et al (2009) Diversities of coral-associated bacteria differ with location, but not species, for three acroporid corals on the Great Barrier Reef. FEMS Microbiol Ecol 68:152–163

Love GD, Grosjean E, Stalvies C, Fike DA, Grotzinger JP, Bradley AS, Kelly AE, Bhatia M, Meredith W, Snape CE, Bowring SA, Condon DJ, Summons RE, Love GD, Grosjean E, Stalvies C, Fike DA, Grotzinger JP, Bradley AS, Kelly AE, Bhatia M, Meredith W, Snape CE, Bowring SA, Condon DJ, Summons RE (2009) Fossil steroids record the appearance of Demospongiae during the Cryogenian period. Nature 457:718–721

Lozupone CA, Stombaugh JI, Gordon JI, Jansson JK, Knight R (2012) Diversity, stability and resilience of the human gut microbiota. Nature 489:220–230

Lynch VJ, Nnamani MC et al (2015) Ancient transposable elements transformed the uterine regulatory landscape and transcriptome during the evolution of mammalian pregnancy. Cell Rep 10(4):551–561

Makarova KS, Haft DH, Barrangou R, Brouns SJJ, Charpentier E, Horvath P, Moineau S, Mojica FJM, Wolf YI, Yakunin AF, Oost J, Koonin EV (2011) Evolution and classification of the CRISPR-Cas systems. Nat Rev Microbiol 9:467–477

Margulis L (1981) Symbiosis in cell evolution. W. H. Freeman, New York

Margulis L (1993) Symbiosis in cell evolution, 2nd edn. W. H. Freeman & Co., New York

Margulis L, Sagan D (2001) Marvellous microbes. Resurgence 206:10–12

Márquez LM, Redman RS, Rodriguez RJ, Roossinck MJ (2007) A virus in a fungus in a plant: three-way symbiosis required for thermal tolerance. Science 315:513–515

Martínez DE, Iñiguez AR, Percell KM, Willner JB, Signorovitch J, Campbell RD (2010) Phylogeny and biogeography of Hydra (Cnidaria: Hydridae) using mitochondrial and nuclear DNA sequences. Mol Phylogenet Evol 57:403–410

May RM, Levin SA, Sugihara G (2008) Complex systems: ecology for bankers. Nature 451:893–895

Mazmanian SK, Liu CH, Tzianabos AO, Kasper DL (2005) An immunomodulatory molecule of symbiotic bacteria directs maturation of the host immune system. Cell 122(1):107–118

Mazmanian SK, Round JL, Kasper DLA (2008) Microbial symbiosis factor prevents intestinal inflammatory disease. Nature 453(7195):620–625

McAuley PJ (1981) Control of cell division of the intracellular Chlorella symbionts in green Hydra. J Cell Sci 47:197–206

McAuley PJ (1985) Regulation of numbers of symbiotic Chlorella in digestive cells of green hydra. Endocyt Cell Res 2:179–190

McCutcheon JP, Von Dohlen CD (2011) An interdependent metabolic patchwork in the nested symbiosis of mealybugs. Curr Biol 21:1366–1372

McFall-Ngai MJ (2002a) Unseen forces: the influence of bacteria on animal development. Dev Biol 242:1–14

McFall-Ngai MJ (2002b) Unseen forces: the influence of bacteria on animal development. Develop Biol 242:1–14

McFall-Ngai M (2008) Are biologists in 'future shock'? Symbiosis integrates biology across domains. Nat Rev Microbiol 6(10):789–792

McFall-Ngai MJ, Ruby EG (1991) Symbiont recognition and subsequent morphogenesis as early events in an animal-bacterial mutualism. Science 254(5037):1491–1494

McFall-Ngai M, Heath-Heckman EAC, Gillette AA, Peyer SM, Harvie EA (2012) The secret languages of coevolved symbioses: insights from the Euprymna scolopes-Vibrio fischeri symbiosis. Semin Immunol 24(1):3–8

McFall-Ngai M, Hadfield M, Bosch T, Carey H, Domazet-Loso T, Douglas A, Dubilier N, Eberl G, Fukami T, Gilbert S, Hentschel U, King N, Kjelleberg S, Knoll A, Kremer N, Mazmanian S, Metcalf J, Nealson K, Pierce N, Rawls J, Reid A, Ruby E, Rumpho M, Sanders J, Tautz D, Wernegreen J (2013) Animals in a bacterial world, a new imperative for the life sciences. Proc Natl Acad Sci U S A 110(9):3229–3236

Miller DJ, Ball EE (2008) Cryptic complexity captured: the Nematostella genome reveals its secrets. Trends Genet 24:1–4

Miller DJ, Hemmrich G, Ball EE, Hayward DC, Khalturin K, Funayama N et al (2007) The innate immune repertoire in cnidaria–ancestral complexity and stochastic gene loss. Genome Biol 8:R59

Minot S, Sinha R, Chen J, Li H, Keilbaugh SA, Wu GD, Lewis JD, Bushman FD (2011) The human gut virome: inter-individual variation and dynamic response to diet. Genome Res 21:1616–1625

Möbius KA (1877) Die Auster und die Austernwirtschaft. Wiegandt, Hempel & Parey, Berlin

Moran NA (2007) Symbiosis as an adaptive process and source of phenotypic complexity. Proc Natl Acad Sci 104:8627–8633

Moran NA, Yun Y (2015) Experimental replacement of an obligate insect symbiont. Proc Natl Acad Sci USA 112:2093–2096

Moran NA, Degnan PH, Santos SR, Dunbar HE, Ochman H (2005) The players in a mutualistic symbiosis: insects, bacteria, viruses, and virulence genes. Proc Natl Acad Sci U S A 102(47):16919–16926

Morgan XC, Segata N, Huttenhower C (2013) Biodiversity and functional genomics in the human microbiome. Trends Genet 29(1):51–58

Moroz LL et al (2014) The ctenophore genome and the evolutionary origins of neural systems. Nature 510:109–114

Mortzfeld B, Urbanski S, Reitzel AM, Künzel S, Technau U, Fraune S (2015) Response of bacterial colonization in Nematostella vectensis to development, environment and biogeography. Environ Microbiol. doi:10.1111/1462-2920.12926

Muscatine L, Lenhoff HM (1963) Symbiosis: on the role of algae symbiotic with hydra. Science 142(3594):956–958

Muscatine L, Lenhoff HM (1965a) Symbiosis of hydra and algae. I effects of some environmental cations on growth of symbiotic and aposymbiotic hydra. Biol Bull 128:415–424

Muscatine L, Lenhoff HM (1965b) Symbiosis of hydra and algae. II effects of limited food and starvation on growth of symbiotic and aposymbiotic hydra. Biol Bull 129:316–328

Muscatine L, McAuley PJ (1983) Transmission of symbiotic algae to eggs of green Hydra. Cytobios 33:111–124

Nawrocki AM, Collins AG, Hirano YM, Schuchert P, Cartwright P (2013) Phylogenetic placement of Hydra and relationships within Aplanulata (Cnidaria: Hydrozoa). Mol Phylogenet Evol 67:60–71

Nebel A, Bosch TCG (2012) Evolution of human longevity: lessons from Hydra (Editorial). Aging 4(11):730–731

Nichols SA, Dayel MJ, King N (2009) Genomic, phylogenetic, and cell biological insights into metazoan origins. In: Littlewood DTJ, Telford MJ (eds) Animal evolution: genomes, fossils, and trees. Oxford University Press, Oxford

Nilsson DE (2013) Eye evolution and its functional basis. Vis Neurosci 30:5–20

Nosenko T, Schreiber F, Adamska M, Adamski M, Eitel M, Hammel J, Maldonado M, Müller WE, Nickel M, Schierwater B, Vacelet J, Wiens M, Wörheide G (2013) Deep metazoan phylogeny: when different genes tell different stories. Mol Phylogenet Evol 67(1):223–233

O'Brien TL (1982) Inhibition of vacuolar membrane fusion by intracellular symbiotic algae in Hydra viridis (Florida strain). J Exp Zool 223(3):211–218

O'Hara AM, Shanahan F (2006) The gut flora as a forgotten organ. EMBO Rep 7(7):688–693

Ochman H, Worobey M, Kuo CH, Ndjango JB, Peeters M, Hahn BH, Hugenholtz P (2010) Evolutionary relationships of wild hominids recapitulated by gut microbial communities. PLoS Biol 8:e1000546

Oliver KM, Degnan PH, Hunter MS, Moran NA (2009) Bacteriophages encode factors required for protection in a symbiotic mutualism. Science 325:992–994

Olson ND et al (2009) Diazotrophic bacteria associated with Hawaiian *Montipora* corals: diversity and abundance in correlation with symbiotic dinoflagellates. J Exp Marine Biol Ecol 371:140–146

Pandolfi JM et al (2011) Projecting coral reef futures under global warming and ocean acidification. Science 333:418–422

Pardy RL (1983) Preparing aposymbiotic hydra. In: Lenhoff HM (ed) Hydra: research methods. Plenum Press, New York, pp 394–395

Piatigorsky J, Kozmik Z (2004) Cubozoan jellyfish: an Evo/Devo model for eyes and other sensory systems. Int J Dev Biol 48:719–729

Plachetzki DC, Oakley TH (2007) Key transitions during the evolution of animal phototransduction: novelty, "tree-thinking", co-option, and co-duplication. Integr Comp Biol 47:759–769

Pochon X et al (2006) Molecular phylogeny, evolutionary rates and divergence timing of the symbiotic dinoflagellate genus *Symbiodinium*. Mol Phylogenet Evol 38:20–30

Putnam NH, Srivastava M, Hellsten U, Dirks B, Chapman J, Salamov A et al (2007) Sea anemone genome reveals ancestral eumetazoan gene repertoire and genomic organization. Science 317:86–94

Rädecker N et al (2015) Nitrogen cycling in corals: the key to understanding holobiont functioning? Trends Microbiol. doi:10.10.16/j.tim.2015.03.008

Rahat M (1985) Competition between chlorellae in chimeric infections of Hydra viridis: the evolution of a stable symbiosis. J Cell Sci 77:87–92

Rahat M, Reich V (1983) A comparative study of tentacle regeneration and number in symbiotic and aposymbiotic Hydra viridis: effect of Zoochlorellae. J Exp Zool 227:63–68

Rahat M, Reich V (1984) Intracellular infection of aposymbiotic Hydra viridis by a foreign free-living Chlorella sp.: initiation of a stable symbiosis. J Cell Sci 65:265–277

Raina J-B et al (2010) Do the organic sulfur compounds DMSP and DMS drive coral microbial associations? Trends Microbiol 18:101–108

Raina J-B et al (2013) DMSP biosynthesis by an animal and its role in coral thermal stress response. Nature 502:677–680

Ralph PJ et al (2001) Zooxanthellae expelled from bleached corals at 33oC are photosynthetically competent. Mar Ecol Prog Ser 220:163–168

Rawls JF, Samuel BS, Gordon JI (2004) Gnotobiotic zebrafish reveal evolutionarily conserved responses to the gut microbiota. Proc Natl Acad Sci U S A 101(13):4596–4601

Rawls JF, Mahowald MA, Ley RE, Gordon JI (2006) Reciprocal gut microbiota transplants from zebrafish and mice to germ-free recipients reveal host habitat selection. Cell 127(2):423–433

Relman DA, Falkow S (2001) The meaning and impact of the human genome sequence for microbiology. Trends Microbiol 9:206–208

Renault S, Stasiak K, Federici B, Bigot Y (2005) Commensal and mutualistic relationships of reoviruses with their parasitoid wasp hosts. J Insect Physiol 51:137–148

Reshef L, Koren O, Loya Y, Zilber-Rosenberg I, Rosenberg E (2006) The coral probiotic hypothesis. Environ Microbiol 8:2,067–2,073

Reyes A, Semenkovich NP, Whiteson K, Rohwer F, Gordon JI (2012) Going viral: next-generation sequencing applied to phage populations in the human gut. Nat Rev Microbiol 10(9):607–617

Reyes A, Wu M, McNulty NP, Rohwer FL, Gordon JI (2013) Gnotobiotic mouse model of phage-bacterial host dynamics in the human gut. Proc Natl Acad Sci U S A 110(50):20236–20241

Richter DJ, King N (2013) The genomic and cellular foundations of animal origins. Ann Rev Genet 47:509–537

Rohwer F et al (2002) Diversity and distribution of coral-associated bacteria. Mar Ecol Prog Ser 243:1–10

Roossinck MJ (2011) The good viruses: viral mutualistic symbioses. Nat Rev Microbiol 9:99–108

Rosenberg E, Falkowitz L (2004) The Vibrio shiloi/Oculina patagonica model system of coral bleaching. Annu Rev Microbiol 58:143–159

Rosenberg E, Zilber-Rosenberg I (2011) Symbiosis and development: the hologenome concept. Birth Defects Res C Embryo Today 93(1):56–66

Rosenberg E, Kellogg A, Rohwer F (2007a) Coral microbiology. Oceanography 20(2):146–154

Rosenberg E, Koren O, Reshef L, Efrony R, Zilber-Rosenberg I (2007b) The role of microorganisms in coral health, disease and evolution. Nat Rev Microbiol 5:355–362

Rosenberg E, Sharon G, Zilber-Rosenburg I (2009) Opinion: the hologenome theory of evolution contains Lamarckian aspects within a Darwinian framework. Environ Microbiol 11(12): 2959–2962

Roux S, Enault F, Robin A, Ravet V, Personnic S, Theil S, Colombet J, Sime-Ngando T, Debroas D (2012) Assessing the diversity and specificity of two freshwater viral communities through metagenomics. PLoS One 7(3):e33641

Ryan JF, Mazza ME, Pang K, Matus DQ, Baxevanis AD, Martindale MQ et al (2007) Pre-bilaterian origins of the Hox cluster and the Hox code: evidence from the sea anemone, Nematostella vectensis. PLoS One 2:e153

Sagan L (1967) On the origin of mitosing cells. J Theor Biol 14(3):255–274

Schmitt S, Tsai P, Bell J, Fromont J, Ilan M, Lindquist N et al (2012) Assessing the complex sponge microbiota: core, variable and species-specific bacterial communities in marine sponges. ISME J 6(3):564–576

Schwentner M, Bosch TCG (2015) Revisiting the age, evolutionary history and species level diversity of the genus Hydra (Cnidaria: Hydrozoa). Mol Phylogenet Evol 91:41–55

Seehausen O, Butlin RK, Keller I, Wagner CE, Boughman JW, Hohenlohe PA et al (2014) Genomics and the origin of species. Nat Rev Genet 15(3):176–192

Sharon G, Segal D, Ringo JM, Hefetz A, Zilber-Rosenberg I, Rosenberg E (2010) Commensal bacteria play a role in mating preference of Drosophila melanogaster. Proc Natl Acad Sci U S A 107(46):20051–20056

Sharp KH et al (2012) Diversity and dynamics of bacterial communities in early life history stages of the Caribbean coral *Porites astreoides*. ISME J 6:790–801

Shinzato C, Shoguchi E, Kawashima T, Hamada M, Hisata K, Tanaka M et al (2011) Using the Acropora digitifera genome to understand coral responses to environmental change. Nature 476:320–323

Shinzato C, Inoue M, Kusakabe M (2014) A snapshot of a coral "holobiont": a transcriptome assembly of the scleractinian coral, Porites, captures a wide variety of genes from both the host and symbiotic zooxanthellae. PLoS One 9:e85182

Shnit-Orland M, Kushmaro A (2009) Coral mucus-associated bacteria: a possible first line of defense. FEMS Microbiol Ecol 67:371–380

Shoguchi E, Shinzato C, Kawashima T, Gyoja F, Mungpakdee S, Koyanagi R, Takeuchi T, Hisata K, Tanaka M, Fujiwara M, Hamada M, Seidi A, Fujie M, Usami T, Goto H, Yamasaki S, Arakaki N, Suzuki Y, Sugano S, Toyoda A, Kuroki Y, Fujiyama A, Medina M, Coffroth MA, Bhattacharya D, Sato N (2013) Draft assembly of the symbiodinium minutum nuclear genome reveals dinoflagellate gene structure. Curr Biol 23:1399–1408

Siegl A et al (2011) Single-cell genomics reveals the lifestyle of Poribacteria, a candidate phylum symbiotically associated with marine sponges. ISME J 5:61–70

Silverman J et al (2009) Coral reefs may start dissolving when atmospheric CO2 doubles. Geophys Res Lett 26:1–5

Silverstein RN et al (2012) Specificity is rarely absolute in coral-algal symbiosis: implications for coral response to climate change. Proc R Soc Ser B 279:2609–2618

Smith CL et al (2014) Novel cell types, neurosecretory cells and body plan of the early-diverging metazoan Trichoplax adhaerens. Curr Biol 24:1565–1572

Sokolov S (2009) Effects of a changing climate on the dynamics of coral infectious disease: a review of the evidence. Dis Aquat Organ 87(1–2):5–18

Srivastava M et al (2008) The Trichoplax genome and the nature of placozoans. Nature 454:955–960

Srivastava M, Simakov O, Chapman J, Fahey B, Gauthier MEA, Mitros T, Richards GS, Conaco C, Dacre M, Hellsten U, Larroux C, Putnam NH, Stanke M, Adamska M, Darling A, Degnan SM, Oakley TH, Plachetzki DC, Zhai Y, Adamski M, Calcino A, Cummins SF, Goodstein DM, Harris C, Jackson DJ, Leys SP, Shu S, Woodcroft BJ, Vervoort M, Kosik KS, Manning G, Degnan BM, Rokhsar DS (2010) The Amphimedon queenslandica genome and the evolution of animal complexity. Nature 466(7307):720–726

Stappenbeck TS, Hooper LV, Gordon JI (2002) Developmental regulation of intestinal angiogenesis by indigenous microbes via Paneth cells. Proc Natl Acad Sci USA 99:15451–15455

Stefanik DJ, Friedman L, Finnerty JF (2013) Collecting, rearing, spawning and inducing regeneration of the starlet sea anemone, Nematostella vectensis. Nat Protoc 8:916–923

Steinmetz PR et al (2012) Independent evolution of striated muscles in cnidarians and bilaterians. Nature 487:231–234

Stern A, Sorek R (2012) The phage-host arms-race: shaping the evolution of microbes. Bioessays 33(1):43–51

Stilling RM, Bordenstein SR, Dinan TG, Cryan JF (2014) Friends with social benefits: host-microbe interactions as a driver of brain evolution and development? Front Cell Infect Microbiol 4:17

Stolarski J et al (2011) The ancient evolutionary origins of Scleractinia revealed by azooxanthellate corals. BMC Evol Biol 11:316

Suga H, Schmid V, Gehring WJ (2008) Evolution and functional diversity of jellyfish opsins. Curr Biol 18:51–55

Suga H et al (2013) The Capsaspora genome reveals a complex unicellular prehistory of animals. Nat Commun 4:2325

Sussman M, Loya Y, Fine M, Rosenberg E (2003) The marine fireworm Hermodice carunculata is a winter reservoir and spring-summer vector for the coral-bleaching pathogen Vibrio shiloi. Environ Microbiol 5:250–255

Suttle CA (2007) Marine viruses – major players in the global ecosystem. Nat Rev Microbiol 5(10):801–812

Tauber AI (2008) Expanding immunology: defense versus ecological perspectives. Perspect Biol Med 51:270–284

Taylor LH, Latham SM, Woolhouse ME (2001) Risk factors for human disease emergence. Philos Trans R Soc Lond B Biol Sci 356(1411):983–989

Tchernov D et al (2004) Membrane lipids of symbiotic algae are diagnostic sensitivity to thermal bleaching in corals. Proc Natl Acad Sci U S A 101:13531–13535

Technau U et al (2005) Maintenance of ancestral genetic complexity and non-metazoan genes in two basal cnidarians. Trends Genet 21:633–639

The Human Micorbiome Project Consortium (2012a) A framework for human microbiome research. Nature 486:215–221

The Human Micorbiome Project Consortium (2012b) Structure, function and diversity of the healthy human microbiome. Nature 486:207–214

Thompson FL, Barash Y, Sawabe T, Sharon G, Swings J, Rosenberg E (2006) Thalassomonas loyana sp. nov., a causative agent of the white plague-like disease of corals on the Eilat coral reef. Int J Syst Evol Microbiol 56(Pt 2):365–368

Thorington G, Margulis L (1981) Hydra viridis: transfer of metabolites between Hydra and symbiotic algae. Biol Bull 160:175–188

Thornhill DJ et al (2013) Host-specialist lineages dominate the adaptive radiation of reef coral endosymbionts. Evolution 68:352–367

Trembley A (1744) Mémoires, Pour Servir à l'Histoire d'un Genre de Polypes d'Eau Douce, à Bras en Frome de Cornes. Verbeek, Leiden

Turnbaugh PJ, Ley RE, Hamady M, Fraser-Liggett CM, Knight R, Gordon JI (2007) The human microbiome project. Nature 449:804–810

Umbach JL, Kramer MF, Jurak I, Karnowski HW, Coen DM, Cullen BR (2008) MicroRNAs expressed by herpes simplex virus 1 during latent infection regulate viral mRNAs. Nature 454:780–783

Umesaki Y (1984) Immunohistochemical and biochemical demonstration of the change in glycolipid composition of the intestinal epithelial cell surface in mice in relation to epithelial cell differentiation and bacterial association. J Histochem Cytochem 32(3):299–304

Upchurch P (2008) Gondwanan break-up: legacies of a lost world? Trends Ecol Evol 23:229–236

Ushijima B et al (2014) Vibrio coralliilyticus strain OCN008 is an etiological agent of acute *Montipora* white syndrome. Appl Environ Microbiol 80:2102–2109

Vallina SM, Simó R (2007) Strong relationship between DMS and the solar radiation dose over the global surface ocean. Science 315:506–508

van Baarlen P, Kleerebezem M, Wells JM (2013) Omics approaches to study host-microbiota interactions. Curr Opin Microbiol 16(3):270–277 van Opstal EJ, Bordenstein SR (2015) Rethinking heritability of the microbiome. Science 349(6253):1172–1173

Van Soest RWM et al (2012) Global diversity of sponges (Porifera). PLoS One 7:e35105

Vega Thurber R, Willner-Hall D, Rodriguez-Mueller B, Desnues C, Edwards RA, Angly F, Dinsdale E, Kelly L, Rohwer F (2009) Metagenomic analysis of stressed coral holobionts. Environ Microbiol 211(8):2148–2163

Verbeken G, Huys I, Pirnay JP, Jennes S, Chanishvili N, Scheres J, Górski A, De Vos D, Ceulemans C (2014) Taking bacteriophage therapy seriously: a moral argument. Biomed Res Int 6:213–216

Vidal-Dupiol J et al (2011) Innate immune responses of a scleractinian coral to vibriosis. J Biol Chem 286:22688–22698

Virgin HW, Wherry EJ, Ahmed R (2009) Redefining chronic viral infection. Cell 138(1):30–50

Waggoner B, Collins AG (2004) Reducto ad absurdum: testing the evolutionary relationships of ediacaran and paleozoic problematic fossils using molecular divergence dates. J Paleont 78:51–61

Wagner GP, Kin K, Muglia L, Pavlicev M (2014) Evolution of mammalian pregnancy and the origin of the decidual stromal cell. Int J Dev Biol 58:117–126

Webster NS et al (2013) Near-future ocean acidification causes differences in microbial associations within diverse coral reef taxa. Environ Microbiol Rep 5:243–251

Weinstock GM (2012) Genomic approaches to studying the human microbiota. Nature 489:250–256

Weitz JS, Poisot T, Meyer JR, Flores CO, Valverde S, Sullivan MB, Hochberg ME (2013) Phage-bacteria infection networks. Trends Microbiol 21(2):82–91

Westra ER, Swarts DC, Staals RH, Jore MM, Brouns SJ, van der Oost J (2012) The CRISPRs, they are a-changin': how prokaryotes generate adaptive immunity. Annu Rev Genet 46:311–339

Wheeler WM (1928) The social insects, their origin and evolution. Harcourt Brace, New York

Williams GC (1994) Biotic diversity, biogeography and phylogeny of pennatulacean octocorals associated with coral reefs on the Indo-Pacific. Proc Seventh Int Coral Reef Symp 2:729–735

Williams GC (2011) The global diversity of sea pens (Cnidaria: Octocorallia: Pennatulacea). PLoS One 6:e22747

Williams GP, Babu S, Ravikumar S, Kathiresan K, Prathap SA, Chinnapparaj S, Marian MP, Alikhan SL (2007) Antimicrobial activity of tissue and associated bacteria from benthic sea anemone *Stichodactyla haddoni* against microbial pathogens. J Environ Biol 28:789–793

Wittlieb J, Khalturin K, Lohmann JU, Anton-Erxleben F, Bosch TCG (2006) Transgenic Hydra allow in vivo tracking of individual stem cells during morphogenesis. Proc Natl Acad Sci USA 103(16):6208–6211

Woese CR (2004) A new biology for a new century. Microbiol Mol Biol Rev 68(2):173–186

Woese CR, Fox GE (1977) Phylogenetic structure of the prokaryotic domain: the primary kingdoms. Proc Natl Acad Sci U S A 74(11):5088–5090

Woese CR, Kandler O, Wheelis ML (1990) Towards a natural system of organisms: proposal for the domains Archaea, Bacteria, and Eucarya. Proc Natl Acad Sci U S A 87(12):4576–4579

Xiao X, Laflamme M (2009) On the eve of animal radiation: phylogeny, ecology and evolution of the Ediacara biota. Trends Ecol Evol 24:31–40

Xiao S, Yuan X, Knoll AH (2000) Eumetazoan fossils in terminal Proterozoic phosphorites? Proc Natl Acad Sci U S A 97:13684–13689

Young GA, Hagadorn JW (2010) The fossil record of cnidarian medusa. Palaeoworld 19(2010):212–221

Zhang X, Sun Y, Bao J, He F, Xu X, Qi S (2012) Phylogenetic survey and antimicrobial activity of culturable microorganisms associated with the South China Sea black coral Antipathes dichotoma. FEMS Microbiol 336(2):122–130

Zhaxybayeva O, Doolittle WF (2011) Lateral gene transfer. Curr Biol 21(7):R242–R246

Zilber-Rosenberg I, Rosenberg E (2008) Role of microorganisms in the evolution of animals and plants: the hologenome theory of evolution. FEMS Microbiol Lett 32:723–735

Index

A
Animal life diversity
  characteristics, 42–43
  closest unicellular relatives, extant animals, 43–44
  cnidarians, 32–34
  ctenophora, 39–40
  early diverging, animal phyla
    Cambrian explosion, 28
    Ediacaran fossils, 28
    evolutionary origins, animal groups, 32
    extant forms *vs.* Ediacaran fossils, 31
    metazoan fossils, 29
    quilted organisms, 30
    Vendobiota, 30
  eyes, 42–43
  muscles, 42–43
  nervous systems, 42–43
  placozoans, 40–41
  sponges, 34–38
  superphyla, 27–28
Antimicrobial peptides function, 91–93
Autism and gut bacteria, 4

B
Bacteria
  bleaching, dysbiosis in corals, 116
  evolution, earth, 12
Bilaterality, 35
Biosphere transformation, 13–14
Bleaching, dysbiosis in corals
  bacteria, 116
  coral-associated microbial consortia under stress, 117–118
  coral disease and opportunistic pathogens, 116–117
  corals, defined, 5, 6
  stress sensitivity, 115
  symbiodinium, 118–119

C
Calcareous spicules, 38
Calcareous sponges, 38
Cambrian explosion, 28
Chlorella vs. Hydra, symbiotic interactions, 93–95
Choanoflagellates, 37
Climate
  change and corals, 123
  via sulfur metabolites, corals, 108–109
Cnidarians, 32–34. *See also Hydra*
  complex immune responses, 57–59
  immunity role, 63–64
  inter-partner signaling pathway, 60–61
  mutualisms and animal–microbe cooperation, 61–63
Coevolution, 49
  animal communities, complexity, 6–8
  phylosymbiosis and, 48–51
Comb jellies, 39–40
Complex immune responses, cnidarians, 57–59
Corals
  bacteria location, 102–103
  bleaching, defined, 5, 6
  and climate change, 123
  climate via sulfur metabolites, 108–109
  components of, 104–106
  coral–microbe interactions, 100–101
  disease and opportunistic pathogens, 116–117
  microbial communities, complexity, 101–102
  microbial consortia under stress, 117–118

Corals (cont)
  nitrogen-fixing bacteria, 107
  ocean acidification, 122
  persistence, 120–122
  probiotic microbes and antimicrobial peptides, 107–108
  reef building, 99
  thermal stress/CO_2 elevation, direct impact, 123
  transmission mode and ontogeny, 103–104
CRISPR/CAS System, 19–20
Ctenophora, 39–40

D

Demosponges, 38
Dysbiosis in corals. *See* Bleaching, dysbiosis in corals

E

Ediacaran
  and Cambrian boundary, 13–14
  vs. extant forms, 31
  fossils, 28
Embryo protection, *Hydra*, 87–91
Evolutionary origins, animal groups, 32
Evolution, earth
  bacteria, 12
  biosphere transformation, 13–14
  CRISPR/CAS System, 19–20
  endosymbiotic theory, 13
  eukaryotic cells, 13
  events in, 14
  genome sequences
    bacterial ancestry in, 15–16
    metazoans, 15–16
  host–microbe interactions, 18
  metazoans
    animal-specific genes, 22–24
    genome sequences, 15–16
  microbes, 11
  multicellularity, cooperation of cells, 21–22
  origins of complexity, 20–21
  plants and animals in, 12
  prokaryotic cells, 13
Extant animals, closest unicellular relatives, 43–44

F

FoxO factors, 85

G

Genome sequences
  bacterial ancestry in, 15–16
  metazoans, 15–16

H

Holobionts
  in changing environmental conditions, 73
  defined, 1, 2
Hologenome
  simplified diagram of, 74
  theory, 74
Homoscleromorpha, 38
Host–microbe interactions
  earth evolution, 18
  Hydra, 80–82
Hydra, 50–51
  antimicrobial peptides function, 91–93
  vs. Chlorella, symbiotic interactions, 93–95
  diversification, 54
  embryo protection, 87–91
  female polyp, 89
  FoxO factors, 85
  host–microbe interactions rationale, 80–82
  Hydra magnipapillata genome, 24
  lineages identification, 53
  maternal–zygotic transition (MZT), 88
  microbiota, 82–84
  model, developmental biology, 79
  mucus layer, spatial host–microbial segregation, 86–87
  species-specific microbiota, 50
  tissue homeostasis, 84–86
  viruses in, 129–131

I

Immunity, cnidarians, 63–64
Infectious disease
  and gut bacteria, 4
  incidence, 2, 3
Infusoria, 37
Inter-partner signaling pathway, 60–61

M

Maternal–zygotic transition (MZT), 88
Metazoans
  animal-specific genes, 22–24
  ecosystems, 135–136
  fossils, 29

genome sequences, 15–16
modularity, power of, 137–138
Microbe-associated molecular patterns (MAMPs), 57
Microbes, 11
   corals
      communities, complexity, 101–102
      consortia under stress, 117–118
   host functional entity, symbionts, 8–9
Microbiota
   diversification, phylosymbiosis, 51–54
   Hydra, 82–84
   incidence, 2, 3
Mucus layer, spatial host–microbial segregation, 86–87
Multicellularity, cooperation of cells, 21–22
Mutualisms and animal–microbe cooperation, 61–63
MyD88, 59

N
Nitrogen-fixing bacteria, corals, 107
NOD-like receptors (NLRs), 58

O
Ocean acidification and corals, 122
Origins of complexity, 20–21

P
Phylosymbiosis
   animal life and fitness, 47–48
   and coevolution, 48–51
   microbiota diversification, 51–54
Placozoans, 40–41
Probiotic microbes and antimicrobial peptides, 107–108
Prokaryotic cells, 13

Q
Quilted organisms, 30

R
Reef building, 99. *See also* Corals

S
Salpingoeca rosetta choanoflagellates, 24
Spatial host–microbial segregation, 86–87
Species-specific microbiota, 50
Sponges
   animal life diversity, 34–38
   calcareous, 38
   demosponges, 38
Stress
   corals, microbial consortia under, 117–118
   sensitivity, corals, 115
   thermal/CO_2 elevation, direct impact, 123
Symbiodinium, 5–6
   bleaching, dysbiosis in corals, 118–119
Symbionts, evolutionary processes
   adaptation to environmental conditions, 73–75
   developmental symbiosis, 68–69
   in evolutionary processes, 69–70
   microbes, as forgotten organ, 67–68
   Nematostella, host–microbe interactions, 71–73
   in speciation, 75–76
Symbiosis, defined, 1

T
Thermal stress/CO_2 elevation, direct impact, 123
Tissue homeostasis, 84–86
Toll/Toll-like receptor (TLR) pathway, 57

V
Vendobiota, 30
Viruses
   bacteriophage therapy in, 131–132
   beneficial types, 128–129
   genetic diversity impact, 127
   in *Hydra*, 129–131
   metaorganisms, 127

Z
Zooxanthellae, 5

Printed by Printforce, the Netherlands